Expert Systems

Expert Systems

Introduction to First and Second Generation and Hybrid Knowledge Based Systems

Chris
NIKOLOPOULOS
Bradley University
Peoria, Illinois

CRC Press
Taylor & Francis Group
Boca Raton London New York

CRC Press is an imprint of the
Taylor & Francis Group, an **informa** business

CRC Press
Taylor & Francis Group
6000 Broken Sound Parkway NW, Suite 300
Boca Raton, FL 33487-2742

First issued in paperback 2019

ISBN-13: 978-0-367-40108-5

Visit the Taylor & Francis Web site at
http://www.taylorandfrancis.com

and the CRC Press Web site at
http://www.crcpress.com

Library of Congress Cataloging-in-Publication Data

Nikolopoulos, Chris
Expert systems: introduction to first and second generation and hybrid knowledge based systems / Chris Nikolopoulos.
p. cm.
Includes bibliographical references and index.

1. Expert systems (Computer science) I. Title.
QA76.76.E95N54 1997
006.3'3—dc21

96-50429
CIP

To
my mother, Maria
to
Oak,
Basil, and Christian
Panagiotis
and to
the memory of my father, Vasilios

PREFACE

Expert Systems are one of the first commercial successes of AI and their use in industry, business, education, science, law and medicine has increased greatly over the past few years. As demonstrated by the huge number of successfully deployed expert systems around the world, using expert systems can result in decreased costs and immense savings, reduced downtime and increased quality and throughput.

An expert or knowledge based system uses knowledge gained through reasoning and heuristics to solve with a high degree of reliability intractable, complex real world problems. Expert systems are usually used to solve those problems for which algorithmic, polynomial time solutions are not available. In addition to being able to solve knowledge intensive problems that are not easily addressed by conventional software, expert systems provide other highly desirable features which include their declarative, nonprocedural nature, and the ability to provide explanations for their decisions and to work with incomplete or uncertain data.

Development in the field has moved beyond first generation expert systems. In second generation expert systems, the knowledge engineering process is viewed more as a modeling activity that entails the construction of a knowledge level model of the system and instantiation of the model with domain specific knowledge. Second generation expert systems are distinguished from their first generation counterparts in that they organize their knowledge base in modules and multilevel structures. For example, they separate control from domain knowledge, employ hybrid representations by combining multiple representation schemes, supply sophisticated semi-automated and automated learning tools to help with knowledge acquisition, and can provide more satisfactory and comprehensive explanations of their actions as a by-product of treating the knowledge engineering process as a modeling activity.

Another very important trend in expert systems development is the merging of multiple paradigms, including the connectionist and evolutionary paradigm, with the traditional logic based approach. Multiparadigm hybrid expert systems draw on the strengths of each individual approach to render more powerful systems. The integration of paradigms enables the system to solve certain subtasks that are not amenable to solution by the traditional "elicit knowledge-represent knowledge" approach and endows the system with learning capabilities.

These new developments in the field warrant that learners and aspiring expert system developers be aware of the options available and be knowledgeable about not only the symbolic approach but also the knowledge level and modeling approach as well as various areas of machine learning.

This book provides an introduction to the field of expert/knowledge based systems and covers current and emerging trends and research areas. In addition to the traditional areas of reasoning, knowledge acquisition, verification, validation, knowledge representation and uncertainty, this book covers second generation expert systems, hybrid expert systems and machine learning, including the connectionist and evolutionary paradigms.

The book is intended for use in undergraduate or first year graduate level classes on knowledge based systems, as well as for use by computer professionals or engineers who wish to acquire a working knowledge of the field and an understanding of its foundational principles. It is self-contained as much as possible, and the only requirement for thorough understanding is some knowledge of basic probability theory and statistics, boolean logic, and set theory.

There are basically two alternatives for developing an expert system: one can use an expert system shell or a programming environment. This book caters to both approaches. The distinction between shells and programming environments is no longer as clear as it was a few years back. There are many expert system shells that provide sophisticated control languages and interfaces to programming language environments. This gives the developer the option of bypassing the built-in knowledge representation and inference mechanisms of the shell and the ability to custom-make the system to fit the application, instead of trying to fit the application to the shell. An introduction to one of the commercially available expert system shells, EXSYS, is included in Appendix B.

Even though expert system shells are getting better and better, and there is an abundance of shells available, no shell is universally accepted (each shares a portion of the market) and each shell implements a limited number of different knowledge representation and reasoning techniques, ways to handle uncertainty, etc. A programming environment provides more flexibility than an expert system shell, at the expense of requiring programming expertise. In this book, Prolog is used as a tool for demonstrating how various expert system concepts can be implemented in a programming environment. Prolog compilers are widely available and standardized to a certain extent and good public domain versions can be obtained or downloaded through the Internet (see Appendix C). The declarative nature and built-in inference engine of Prolog compilers make it possible to implement expert system concepts more easily than by using other procedural languages or LISP and eliminate the overhead of building an inference engine. A very detailed and self-contained tutorial in Prolog is contained in Appendix A. Readers with previous programming experience in any language will certainly find it much easier to understand the Prolog material if they choose to go through it but the inclusion of Prolog is not central to the book. The instructor or reader who wishes only to address shell-based application development can omit the Prolog based examples and the tutorial on Prolog without any loss in continuity.

I have used preliminary versions of the book in both undergraduate and graduate level courses for computer science majors and undergraduate/graduate courses for engineering and computer information systems majors. Two alternate paths can be taken when using the book for a course, depending on the type of audience.

In a computer science course, the approach mainly uses a programming language environment for expert system development. Prolog is used for implementation of the semester-long project of building a medium- to large-size expert system with modules for handling uncertainty and multiple knowledge representation paradigms. At the same time, the students are exposed to the use of expert system shells and given assignments on implementing small- to medium- size expert systems using a shell. During the course of the semester they are given the assignment of implementing small prototypes in both Prolog and the expert system shell.

For computer information systems and engineering students, Chapter 2 does not need to be covered in detail and Appendix C on Prolog can be omitted. The use of Prolog is deemphasized and an expert system shell is the main tool used for system development. The various sections in the other chapters that are concerned with implementation of various expert system concepts in Prolog can be skipped as well without any loss of understanding, and such sections, marked with an asterisk (*) are optional. Depending on the expert system shell used in the class, some of these concepts (for example, frames or techniques for treating uncertainty) may be supported by the shell and the instructor can substitute the Prolog sections on implementation by the way the particular shell used in class implements the concepts. Assignments are given for building applications by using the techniques supported by the shell (for example, using EXSYS, an assignment could be to create and manipulate a frame based knowledge base, or to customize the default uncertainty mechanism of EXSYS so that it can handle uncertainty using the confidence factors or fuzzy logic approach instead). The term project on expert system development is done using an expert system shell as well.

Exercises and extensive references for further in-depth study are provided at the end of chapters. The exercises fall into one of three categories. The first type does not require programming in Prolog. The second type can be done either in Prolog or using an expert system shell, assuming that the shell to be used supports the intended knowledge representation structure, ways to deal with uncertainty, etc. Finally, some of the exercises are meant to be done in Prolog and are marked with an asterisk.

The contents are organized as follows: Chapter 1 contains an introduction to the general concepts underlying the field of expert systems. Chapter 2 covers theoretical foundations of expert systems in logic programming, inference techniques and nonmonotonic logics. Chapter 3 describes various knowledge representation schemes and Chapter 4 covers knowledge acquisition, verification and validation, second generation expert systems, the knowledge level and the generic tasks and problem solving methods approach. Chapter 5 presents various ways for handling uncertainty in expert systems. Chapter 6 introduces machine learning techniques (including inductive techniques and genetic algorithms) and their use in alleviating the knowledge acquisition bottleneck. Chapter 7 is an introduction to the connectionist approach and neural networks and Chapter 8 shows how the connectionist (neural expert systems), evolutionary (classifier systems) and the symbolic paradigms can be combined into Hybrid Expert Systems. Appendix A provides an introduction to the declarative language Prolog. Appendix B gives an introduction to the expert shell EXSYS and Appendix C gives a listing of expert system shells, available versions of Prolog and a list of links to web sites of related interest.

Many thanks are expressed to my students who endured through preliminary versions of the book. Many of them contributed to making the book better and especially the students in the AI and expert systems classes the author taught in the fall of '95 and spring of '96. I am also indebted to the contributions from reviewers and colleagues who helped improve the book quality. Many thanks to the Marcel Dekker team, especially Editor-in-Chief Russell Dekker and Production Editor Brian Black, for their assistance in completing and improving the manuscript. Finally, I am apologetic to my sons and grateful to my wife for enduring my virtual isolation from family life while writing this book.

Chris Nikolopoulos

CONTENTS

* Discussion specific to Prolog

* Discussion specific to Prolog.

* Discussion specific to Prolog

* Discussion specific to Prolog.

CHAPTER 1

INTRODUCTION TO EXPERT SYSTEMS

1.1 ARTIFICIAL INTELLIGENCE

Artificial Intelligence (AI) is an interdisciplinary field of study that emerged in the 1950s from origins in computing, engineering, psychology, mathematics and cybernetics. The main goal of AI is to build systems that exhibit intelligent behavior and perform complex tasks with a level of competence that is equivalent or superior to the level currently exhibited by human experts. The term Artificial Intelligence (AI) was introduced by McCarthy in 1956 during a conference at Dartmouth College and has since been established as representing the field, even though the term is a little provocative to some and the very idea of "artificial intelligence" seems to raise various philosophical objections (Dreyfus, [3]).

Another term, which could describe this area of study and would probably be less controversial, is the term *Heuristic Complex Problem Solving*. The types of problems that AI techniques aim to solve, often by use of heuristics, are complex problems with ill-defined domains for which no algorithmic solution is known or which belong to the class of NP-complete problems (the known algorithm is of exponential running time and a solution cannot be obtained for large sets of input data due to time considerations). Fertile areas, where AI techniques are particularly suitable due to the nature of the problems encountered in them, include learning, pattern recognition, vision, robotics, natural language understanding and translation. AI techniques have found applications in a host of domains including engineering, medicine, business, education and the sciences.

In the early days of development, researchers pursued the goal of building general problem solvers like the Logic Theorist and GPS (General Problem Solver), (Newell and Simon, [18,19]). Progress in that direction stalled, the developed systems did not find widespread applicability and were unable to provide complete solutions to a large variety of complex problems. As a result a change of focus took place. It was realized that specialized problem solving knowledge could be developed and applied to the various domains, instead of searching for an omnipotent, general purpose, problem solving methodology. Specialized problem solvers and techniques started being developed to solve particular problems. The advent of Expert or Knowledge Based Systems is a result of pursuing this goal of constructing specialized problem solvers.

Two of the most prominent approaches to Artificial Intelligence, which have emerged over the past several years, are the *symbol manipulating approach* and *the connectionist approach. Expert systems (knowledge based systems)* and *artificial neural networks* have emerged from the symbolic and connectionist approaches respectively, as two of the most widely used and commercially successful applications of Artificial Intelligence.

The symbolic approach uses logic for the representation of knowledge, reasoning and deduction. Expert system development entails the conversion of information about a bounded problem domain into knowledge, which is then represented in a format suitable for computer manipulation. The created depository of knowledge, known as the knowledge base, can then be used by various deductive reasoning techniques to derive solutions.

The foundation of the connectionist approach lies with a computational mechanism known as *artificial neural network* (ANN). ANNs are composed of smaller units called "neurons" which are connected to each other to form a network. The neurons are individual units that perform simple local calculations. They cooperate by exchanging their local knowledge in the network structure enabling the ANN to achieve a common global goal. ANNs have the ability to *learn* their behavior from raw data.

Although development in the two areas has mostly progressed independently from each other, it has recently been realized that the merging of traditional expert systems, neural networks and other AI learning paradigms into hybrid systems can give rise to more powerful and efficient systems.

1.2 BASIC EXPERT SYSTEM CONCEPTS

Feigenbaum, ([11]), defined an expert system as "an intelligent computer program that uses knowledge and inference procedures to solve problems that are difficult enough to require significant human expertise for their solution. The knowledge of an expert system consists of facts and heuristics." This definition is basically still valid even though, as we will see below, it is a bit outdated when used to refer to the second generation of expert systems.

An expert system solves problems in a narrow domain of expertise and cannot be a general problem solver. For example, an expert system should not attempt to solve problems in a very broad domain, such as "the medical domain," but instead it would address specific subareas of the medical domain. This is demonstrated by MYCIN and Internist which are expert systems for diagnosing infectious blood diseases and internal medicine problems respectively.

A computer program is not given the label of an expert system just because of its ability to perform like an expert in a domain, as the name "expert system" seems to suggest. For example, many computer programs, which can now compete at the grandmaster level in chess, are based more on a brute force approach and are not necessarily identified as expert systems. It is more the characteristics of a system that define the system as belonging in the class of "expert systems" than simply its performance. These characteristics include the system architecture, the encoding of knowledge in a knowledge base, the separation of domain from control knowledge, the ability to reason under uncertainty, and the availability of explanation facilities, knowledge acquisition tools and user interfaces. Another distinguishing characteristic of expert systems is that they derive solutions by using heuristics rather than the algorithmic approach of conventional programs. An expert system emulates the decision making and problem solving capabilities of a human expert in his/her area of expertise and is expected to perform at a level which is comparable to that of a human domain expert. An expert system should ideally be at least as competent as human experts (in the narrow domain of intended expertise) to be employable in the field. In the case where there is a lack of available human expertise on site and the application is not critical (for example in tutoring systems) even lesser performance may also be acceptable.

The four main architectural components of an expert system are: the *knowledge acquisition module*, the *knowledge base*, the *inference engine* and the *input/output interface*. Figure 1.1 presents the overall general architecture of an expert system.

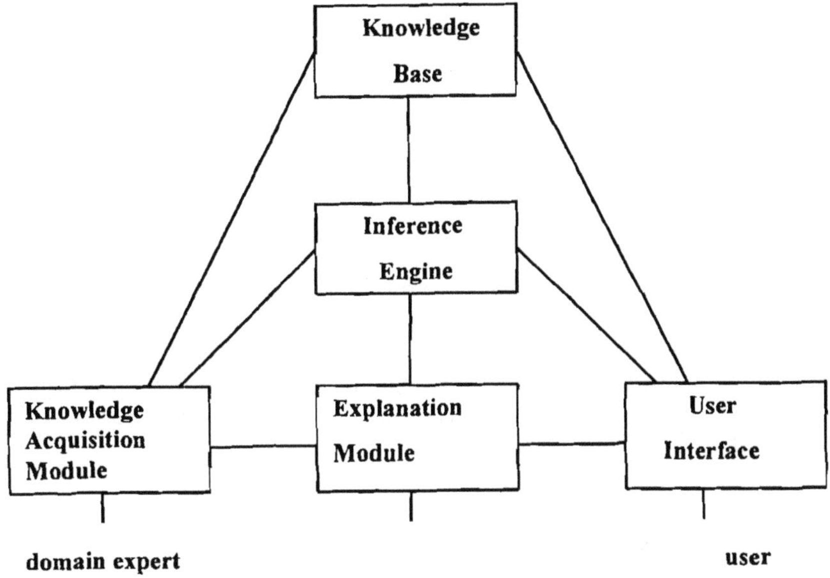

Figure 1.1 The Architecture of an Expert System.

1.2.1 The knowledge base

The *knowledge base* contains the domain specific and control knowledge which is used to solve problems in the domain. This knowledge can be elicited from the experts or it can be learned by the system itself. During the *knowledge elicitation* process, the *knowledge engineer* (system developer) elicits the required problem solving knowledge from the domain experts and other sources, such as books manuals, etc. In automated knowledge acquisition the system itself uses machine learning techniques to directly learn the knowledge from raw data.

The conceptual model of the elicited knowledge is converted in a format suitable for computer manipulation in a process called *knowledge representation*. Knowledge can be represented and stored in the knowledge base in various forms. For example, one of the most commonly used ways to represent knowledge is in the form of If-Then rules. Expert systems that use rules to represent knowledge are called *rule based expert systems*. A rule has the form:

IF (conditions are satisfied)
THEN (take an action or deduce new conclusions).

For example,

IF (temperature rises above 100°) THEN (turn knob #1)

or

IF (temperature is high) THEN (pressure is low)

Knowledge can be represented in a variety of other formalisms as well, including semantic networks, conceptual graphs, frame and object oriented schemes and Petri nets. The topic of knowledge representation is covered in detail in Chapter 3.

1.2.2 The inference engine

The *inference engine* contains general algorithms, which are able to manipulate the knowledge stored in the knowledge base in order to solve problems. The inference engine is separated from the domain knowledge and is general and not domain dependent. The inference mechanism used depends on the knowledge representation chosen since each knowledge representation formalism has its own associated inference techniques. The same inference engine can be used for solving problems in a variety of domains as long as they are modeled using the representation formalism that the inference engine is intended for.

The typical inference engine is based on an *inference rule* (a way to deduce new knowledge from the existing knowledge base) and a *search strategy*. The inference rule must be *sound* and *complete*. As an example, consider the case of rule based systems. One option would be to build the inference engine of a rule based expert system around the principle of resolution. *Resolution* is a complete and sound inference rule that can be applied to derive deductions from a knowledge base of rules and facts. Resolution is *complete* in the sense that if a piece of knowledge is deducible from the knowledge base, resolution can deduce it. It is *sound* in the sense that implications made using resolution are always valid. In general, an inference engine should be based on a deductive mechanism that is sound and complete. Additionally, the inference engine will usually include a *search strategy* in order to most efficiently search for a solution among the vast numbers of possible inference paths. Details on inference rules and the theoretical foundations of expert systems are provided in Chapter 2.

1.2.3 Input/output interface and explanation

The input/output interface defines the way in which the expert system interacts with the user, the environment and other systems such as databases, and spreadsheets. Interfaces are usually graphical (GUI) with screen displays, windowing and mouse control. They display output and receive input either from the user or directly from attached devices. In addition, the interface provides a bridge (hook) to other systems like databases and spreadsheets and enables the user to ask for justification of the system decisions through an explanation facility. Natural language interfaces are also available in some limited domain areas.

The ability of an expert system to explain its recommendations is one of the advantages of expert systems over other approaches (for example, over a neural network that works as a black box). A well designed and effective user interface allows for a good explanation facility which justifies the system's reasoning and can be a very important factor to the eventual successful deployment of an expert system. Explanation in expert systems is addressed in Chapter 4.

1.2.4 Uncertainty in the knowledge base

The knowledge involved in solving complex, real world problems is usually uncertain, imprecise and incomplete. *Imprecision* refers to values which cannot be measured accurately or are vaguely described. *Incompleteness* refers to the unavailability of some or all of the values of an attribute and *uncertainty* is a subjective estimate about the truth value of a fact or the validity of a rule. In addition, *inconsistencies* can be present in a knowledge base, in particular when there is more than one source from where knowledge is elicited.

There are many factors which can contribute to imperfection in the knowledge base. For example, imprecision can be generated by input sensors that often introduce noise to data readings, erroneous or corrupted data, and vagueness (the temperature is "high") in the expert's heuristics. Facts, rule conditions, rule hypotheses or rules themselves can be uncertain. For example, consider the following rule which describes an expert's heuristic problem solving knowledge:

IF (carbon monoxide detector goes off) THEN (there is a gas leak).

The expert may be uncertain of the rule itself (other factors than a gas leak may also cause the alarm to go off) and a reliability factor of 70% may be assigned to the rule. Uncertainty could also be present about whether the alarm went off. It may be only 90% certain that it went off (other devices may also remit a sound similar to the alarm). Under these circumstances, when the rule is applied, the hypothesis "there is a gas leak" will be derived, but not with absolute certainty. How are the uncertainties of the premises and the rule to be combined in order to derive the level of certainty in the conclusion?

Another factor contributing to uncertainty is the use of inherently fuzzy and vague terms during reasoning. For example, consider the rule:

IF (income is high) THEN (approve the loan).

The set of income values that can be classified as "high" is not a crisp set. It would not be very efficient to set a threshold, for example 90,000, and classify an income as "high," if it is greater than the threshold. To do so, would lead to some really unjustifiable decisions. For example, an applicant of yearly income 90,001 would be approved of a loan and an applicant of yearly income equal to 89,999 denied.

Another characteristic of human reasoning that expert systems should be able to emulate is the *non-monotonicity* of thought. Classical logic is monotonic in the sense that when a conclusion is deduced it cannot be withdrawn in the light of new evidence. For example, given the fact that Joe lives 20 miles from campus and the rule of thumb that "if someone lives more than 5 miles from campus then he/she drives to school," a system based on classical logic would deduce that "Joe drives to school." But what if new information becomes available that Joe is blind?

Since human experts are still able to reason and perform their problem solving activities in the presence of imperfect knowledge, expert systems should also be able to accurately model and reason in such domains if they are to be effective. Boolean, monotonic logic is not adequate for solving problems in imperfect domains. Several methodologies have been developed for dealing with imperfect knowledge in expert systems. Methodologies for handling non-monotonicity and the presence of incomplete information in the knowledge base, such as Reiter's default logic and circumscription, are presented in Chapter 2. Chapter 5 provides an introduction to methodologies for reasoning under uncertainty and imprecision. Various techniques for treating uncertainty are presented including the Bayesian, Certainty Factors, Dempster-Shafer, and the Belief Networks approaches. In addition, Chapter 5 covers fuzzy logic and fuzzy inference which generalize boolean and multi-valued logics and allow representation and reasoning under imprecision and vagueness.

1.3 KNOWLEDGE ENGINEERING

The term *Knowledge Engineering* is typically used to describe the expert system development process. A *knowledge engineer* is the individual engaged in knowledge engineering. Expert system development entails the processes of knowledge acquisition, user interface design, selection of software, hardware and implementation. Verification and validation of parts of the expert system and of the system as a whole takes place concurrently to development and also after the system development is complete.

1.3.1 Knowledge elicitation and representation

Knowledge acquisition, (KA), is the process of collecting the knowledge necessary for problem solving in the application domain, and encoding it in a formalism amenable to efficient computer manipulation. Knowledge acquisition can be viewed as the composite of two tasks: knowledge elicitation and knowledge representation.

In *knowledge elicitation*, the domain and problem solving knowledge is gathered and structured in a conceptual model. The knowledge engineer uses various means at his disposal for eliciting the domain and control knowledge, including interviews with the experts, available book and manual references, reports, etc. In addition, various semi-automated and automated techniques have been developed which assist the knowledge engineer with the KA task.

In *knowledge representation*, the acquired knowledge is converted in a form suitable to computer manipulation. The processes of knowledge elicitation and knowledge representation are not necessarily strictly sequential. It is usually the case that knowledge elicitation takes place throughout the lifecycle of development, as it may be realized that the knowledge base is incomplete or inaccurate and may need to be modified.

Chapter 3 provides coverage of various formalisms that can be used for knowledge representation. The topics of knowledge elicitation, verification and validation are covered in Chapter 4.

1.3.2 The knowledge acquisition bottleneck

A distinction can be made between "*cognitive*" and "*subcognitive*" approaches to intelligence. Expert systems founded on logic inference belong to the cognitive approach. On the other hand, human experts routinely use mental activities at the subcognitive level when solving problems. Such problem solving activities are difficult to verbalize or to even become aware of and capture using the cognitive approach to expert systems. This makes the process of extracting the knowledge necessary for problem solving from a human expert extremely tedious, time consuming and difficult. Knowledge acquisition is the major bottleneck in expert system development.

Several knowledge acquisition tools are presently available and they can be generally classified as *conceptualization*, *task specific*, *model based or refinement* tools ([9]). Conceptualization tools, for example ETS and AQUINAS ([2], [3]), are based on theories from psychology like personal construct theory and techniques based on repertory grids. Task specific tools are based on the generic tasks approach to knowledge acquisition. Under this approach some major classes of tasks have been identified called generic tasks, (for example constraint satisfaction, diagnosis or classification tasks, etc.), and task specific problem solving methods have been developed for each type of task. SALT is an example of a task specific tool ([16]). Refinement tools are designed to improve existing knowledge bases, for example by identifying inconsistencies and omissions. TEIRESIAS is a typical example of this type of tool and is used for maintenance of knowledge bases created with the EMYCIN expert system shell, ([5]). Chapter 5 examines various approaches to building semi-automated tools that provide assistance to the knowledge engineer or the expert with the knowledge acquisition process.

Progress has been made in building more flexible systems that can build knowledge structures inductively or learn to improve their own performance without interference from the programmer. Techniques from the subfield of AI called *machine learning* are used to alleviate the knowledge acquisition bottleneck by automating the process of generating knowledge bases. For example, the evolutionary and connectionist approaches, represented by genetic algorithms and neural nets respectively, can be viewed as methodologies that address subcognitive types of knowledge and learning. In fact, the paradigms of genetic algorithms, classifier systems and neural networks are often merged with traditional expert systems to form *Hybrid Expert Systems* that can render an even more powerful problem solving tool than either approach used alone. Genetic algorithms and classifier systems are introduced in Chapter 6 and neural networks in Chapter 7. Various other machine learning techniques are also examined in Chapter 6. Chapter 8 discusses the convergence of the three paradigms into building a hybrid expert system.

1.4 FIRST AND SECOND GENERATION EXPERT SYSTEMS

1.4.1 Evolution from the first to the second generation

Expert system technology has evolved from *first generation* to *second generation expert systems*. Second generation expert systems tend to address some of the weaknesses of first generation expert systems and provide improved explanation facilities, and tools to help with the knowledge acquisition process, reusability and maintainability of expert systems.

First generation expert systems did not distinguish between different types of knowledge, they usually employed a single knowledge representation formalism and they explained their recommendations by rephrasing the rules involved in the inference chain that led to the solution. In the early systems, knowledge acquisition was simply considered to be the process of eliciting knowledge from the expert.

Second generation expert systems are distinguished from their counterparts of the first generation in the following ways. They organize their knowledge base in modules and multilevel structures. They employ hybrid representations by combining multiple representation schemes. They supply sophisticated semi-automated and automated tools to help with knowledge acquisition, and they can provide more satisfactory and comprehensive explanations of their actions as a byproduct of treating the knowledge engineering process as a modeling activity. Second generation expert systems separate control knowledge (for example, ask the general before you ask the specific) from domain specific knowledge and organize the knowledge in multi-level structures. Explanation is viewed as a problem solving activity by itself and the separation of knowledge in many types and levels enables the system to give better explanations at various levels of abstraction depending on the user's level of expertise and comprehension.

Hybrid systems are also becoming commonplace where the connectionist (neural networks) and evolutionary (genetic algorithms and classifier systems) paradigms are combined with the symbolic approach. The integration of paradigms enables the system to solve subtasks that are not amenable to solution by the traditional "elicit knowledge-represent knowledge" approach (such as the ability to learn or improve its knowledge base).

1.4.2 Development as a modeling activity, knowledge level, generic tasks

In Newell's *knowledge level* model, ([18]), the knowledge level contains meta-knowledge about problem solving strategies and how and why control proceeds in a certain way. Many different abstraction levels of meta-knowledge are possible. The separation of domain and control knowledge makes it easier to modify or maintain the knowledge base, has the effect of building better explanation facilities and even helps with the knowledge acquisition process. As mentioned in the previous section, second generation expert systems view the knowledge engineering process more as a modeling activity. Development entails the construction of a knowledge level model of the system and instantiation of the model with the specific domain knowledge.

Expert Systems can be modeled at the knowledge level by describing the domain in terms of tasks (describe "what" is to be done; goals and subgoals of the system), problem solving methods (describe "how" the goals are to be achieved) and domain models (describe the domain specific knowledge required by the methods). Partitioning knowledge in this way makes it easier to acquire, update, reuse and understand.

The *generic tasks* and *problem solving methods* approach to expert system development attempts to characterize the application domain in terms of typical tasks and typical methods used to perform the tasks. An example of a system that places emphasis on the generic task approach is the KADS system, ([25]). In KADS tasks are decomposed into subtasks until inferences can be applied on them.

The generic tasks and methods approach for developing expert systems is in a sense similar to the abstract data types and data structures approach to procedural language programming. An abstract data type (ADT) is a mathematical model of the structure of a set of data elements together with a set of operations that are to be performed on these elements. A data structure is a specific implementation of the ADT. For example, the ADT *list* is defined as a sequence of elements with the set of operations *add_element*, *delete_element*, etc. The ADT *list* can be implemented using a variety of data structures such as one dimensional arrays or linked lists. In second generation expert systems, the ADT approach of organizing data has been transferred over to the organization of knowledge. A task corresponds to an abstract data type and the problem solving methods to the ADT's operations. For example, the generic task *diagnosis* can be defined as: "Given a set of symptoms find the cause." The general method associated with this task is the *cover and differentiate* method that can basically be defined as: "Given a set of symptoms find all the possible causes (*cover*) and then choose the most likely among the possible causes (*differentiate*)." Another method would describe how to find all possible causes (cover), etc. These methods on how to solve problems of diagnostic nature are generic and universal, independent of the application domain. Given a particular domain, if the task to be performed is recognized as a diagnostic task, only the domain specific knowledge about the problem has to be encoded in the predefined representation formalism that was judged as the most appropriate for problems of the diagnostic type.

The approach of generic tasks and problem solving methods promises a partial solution to the knowledge acquisition bottleneck. The structuring of the knowledge into generic task types, problem solving methods and domain specific knowledge has the effect of making the knowledge acquisition process and the maintenance and update of the knowledge base much easier. The knowledge acquisition problem is reduced to being able to identify a specific problem as belonging in one of the classes of generic tasks. Several knowledge acquisition tools specific to different task types have been built which guide the expert system developer though a structured knowledge elicitation for that specific problem type. The generic tasks and problem solving methods approach to knowledge engineering is further explained in Chapter 4.

1.4.3 Second generation expert systems

Explanation facilities were not adequate in first generation expert systems. The lack of separation of control knowledge from domain knowledge made it difficult to provide sufficient explanation. Explanation was primarily based on the retracing of the inferencing steps and was a subset or superset of the chain of deductions. This was inadequate as explanations produced that way tend to be hard to understand, low level and unnecessarily detailed. It was not possible to provide explanations based on the general concepts, strategies and problem solving methods that were implicit in the knowledge base. They would contain terminology internal to the system representation of the knowledge, which was incomprehensible to the user. They also could not provide explanations at different levels of abstraction or explain general problem strategies used. In the second generation of expert systems, meta-knowledge is used to provide an explanation that overcomes the weaknesses given above. The introduction of the "knowledge level" enables the system to accept the user's feedback and if the user does not understand, the system can further explain by changing the detail level or the perspective of explanation. *Control knowledge* ("how") is separated from *associational knowledge* ("why") and *domain specific knowledge* making it easier to explain the system actions without resorting to in order tracing of the rule firings.

Validation and update of knowledge were more difficult in first generation expert systems, as the knowledge was spread throughout the knowledge base. The separation of knowledge into different types makes it easier to validate, verify and update the knowledge base in second generation expert systems.

Second generation expert systems facilitate reusability of knowledge. Some knowledge is generic and can be applied across different domains, but reusability of such parts of the knowledge base was not possible in first generation systems due to the lack of separation of control from domain knowledge.

Knowledge acquisition is more focused and easier in second generation expert systems that use machine learning techniques or the generic tasks and problem solving methods approach.

1.5 ADVANTAGES AND DISADVANTAGES OF EXPERT SYSTEMS

Expert Systems can be viewed as programs that construct models of some real world domain. The models are used to control the domain, perform reasoning, etc. As models, an expert system abstracts a real world problem, only considering important and relevant characteristics (actually what the modeler considers as relevant and important characteristics) and ignoring the rest. It has to be this way, otherwise the program would be lost in the infinitesimal amount of detail that a real world problem entails. For example, consider the problem of navigating a robot around obstacles in a room in order to perform certain tasks. Information about the position of the various obstacles has to be included in the model. The color of the various obstacles, the color of the wall or the number and position of the windows on the wall would not normally be relevant information and would not be included in the model. Including details in the model, which are not necessary for the solution of the immediate goal, will only magnify the so-called *frame problem* in AI and make it from difficult to impossible to keep track of how changes in the modeling environment affect other entities in the environment. An inherent difficulty with modeling the world is of course the problem of deciding what is relevant and what not. If for example, one of the tasks the robot had to perform in the room was to paint the wall, then the color of the wall and the position of the windows would all of the sudden be very relevant.

Another problem inherent in the modeling activity is the difficulty to foresee all the possible conditions that may happen in a real world problem and which can affect the problem solution. Consider a radar system for detecting nuclear attacks that can be spooked into believing an attack is underway by an unaccounted for flock of birds flying over the equipment.

Expert systems as a modeling activity are nonalgorithmic in nature. As a result, they cannot be proven correct. There may be unforeseen circumstances by the designers in which the model will fail. This makes the use of expert systems not very appropriate for real time, critical applications in which a decision has to be made within a very short period of time; the wrong decision can have catastrophic effects and human experts would not have the time to review the system's recommended solution. If we left it up to our hypothetical nuclear attack expert system, the system would shoot back missiles for the just cause of a flock of geese invading the country.

In the early 1970s, the direction of AI research shifted from trying to build general problem solvers, to domain specific problem solvers. As mentioned above, expert systems are a product of this approach and deal with problem solving in a restricted domain. They lack the overall knowledge, general problem solving skills and functional ability of human experts that are needed to adjust and transfer their skills to unforeseen circumstances. The fact that expert systems are not general problem solvers is another of their weaknesses. They do not really know the limits of their expertise and will tend to provide an output even when the case submitted to them falls outside their domain of expertise. For example, a planning expert system will still be trying to find the most efficient flexible manufacturing plan in the mist of a fire in the plant and will not bother to sound the alarm or tell everybody to get out. It only knows how to do one thing, design plans.

Advantages of expert system technology include the ability to solve, with high degree of reliability, intractable, complex problems, for which algorithmic, polynomial time solutions are not available. One of the driving needs for expert system development is to capture critical and scarce expertise and distribute it, throughout the organization. Human expertise is a scarce commodity and expert systems can provide an accessible, available alternative that can also be used as a training tool for beginners. They can replace a human expert or they can provide assistance to the human expert in problem solving. In that capacity, their declarative, nonprocedural nature makes the problem solving activity easier. One highly desirable feature that expert systems provide and conventional software do not is the ability of expert systems to provide explanations for their decisions and to work with incomplete or uncertain data. As demonstrated by the huge number of successfully deployed expert systems around the world, using expert systems can result in decreased costs and immense savings for an organization, reduced downtime and increased quality and throughput.

1.6 EXPERT SYSTEM SHELLS, PROGRAMMING ENVIRONMENTS

The tools available for building expert systems can be classified into *expert system shells* and *programming environments*. An expert system shell is a software system that facilitates expert system development by providing a collection of appropriate functions. An expert system shell usually contains:

1. A set of knowledge representation structures
2. A built in inference engine

3. Knowledge acquisition tools to help the knowledge engineer in the knowledge elicitation process
4. A user interface and explanation facility
5. Interfaces to other software systems (spreadsheets, databases, programming languages etc.)

An expert system shell can greatly facilitate development, but this is often at the expense of flexibility. The developer is limited to using the knowledge representation mechanisms, inference techniques, ways to handle uncertainty and user interfaces that the shell provides. This can result in trying to fit the problem to the shell, rather than customizing the software to fit the problem. A plethora of expert system shells exists in the market today and a partial list of the available shells is given in appendix C. A description of the expert system shell EXSYS is provided in Appendix B. For ways to select an appropriate shell based on desired features as defined by the problem characteristics, the reader is referred to [23]. The majority of expert systems in operation today are PC based and have been built using commercial expert system shells, ([20]).

The use of a programming environment gives the developer greater flexibility, as he/she has total control over the techniques to be used. Of course greater expertise is required and it may be more time consuming to complete an expert systems project using a programming environment rather than a shell. Expert systems have been developed in procedural languages like C or Pascal, in general AI languages like LISP or Prolog and their object oriented extensions CLOS (Common Lisp Object System) and L&O (Logic and Objects), and in specialized production systems languages like OPS5 ([4]), CLIPS ([11]) or COOL (object oriented language extension of CLIPS).

In procedural languages or the list processing language LISP, the programmer can encounter a lot of overhead since he/she has to develop the inference engine and representation formalisms. One advantage of a logic programming environment like Prolog is that it contains a built in inference mechanism, which is easily modifiable. Its declarative nature makes it easy to express the available knowledge in Prolog format. Even though Prolog employs the rule knowledge representation formalism, other representation formalisms can easily be embedded in the Prolog programming environment. Prolog can provide a unified medium of development and the ability to have various representation techniques available in the same environment. A self-contained introduction to Prolog is given in Appendix A.

1.7 EXPERT SYSTEM APPLICATIONS

A proliferation of expert system development has occurred over the last few years in commercial and industrial applications and in such diverse areas as the public sector, business, finance, education, manufacturing, telecommunications, the medical domain and law. We list below a few of the most widely known expert systems. For a more thorough list of applications the reader is referred to [8].

DENTRAL

Dentral, ([14]), was one of the first expert systems to be developed and it used heuristic knowledge in the form of rules to address a combinatorial explosion problem. The system analyzes mass spectroscopy data to identify the molecular structure of unknown compounds. Even though it lacked explanation facilities, it was widely used by chemists. Dentral was written in LISP.

MYCIN

Together with DENTRAL, MYCIN (Shortliffe, [22]) is one of the most influential and widely quoted expert systems, even though it was not completely accepted by users in its application domain. It was designed to perform diagnosis of infectious blood diseases and recommend an appropriate treatment. Its inference engine was based on backward chaining. In order to prove a query, *backward chaining* chooses a rule whose conclusion part matches the query and the search to prove the rule's premises continues in a recursive manner (see chapter 2, section 11). MYCIN rules had *confidence factors* associated to them. A confidence factor is a numeric value indicating the uncertainty about the rule's applicability. Eliminating MYCIN's domain specific knowledge gave rise to the expert system shell EMYCIN. One example of a system developed by EMYCIN is PUFF, an expert system for identifying lung disorders and suggesting treatment. PUFF was able to receive input directly from medical test instruments and use it to produce its diagnosis. It was widely accepted in the medical community. We will refer to MYCIN in various instances in the text as the typical example of first generation expert systems.

R1/XCON

The first major successful commercial application of expert systems. R1/XCON is a *forward chaining* system (if all of the conditions of a rule become valid, the rule fires to ascertain its conclusion as valid, see Chapter 2, Section 11) and it was implemented using the production system language OPS (versions 4 and 5). Inference proceeds in *recognize-*

select-act cycles: all rules whose conditions are satisfied are candidates for activation; a *conflict resolution strategy* chooses which rule should be activated; the chosen rule fires, performing its action part (see Chapter 4, Section 2). R1/XCON is still in use today for computer hardware configuration at DEC (Digital Equipment Corporation). Since its inception it has continually been improved and expanded.

HEARSAY I and II
Hearsay ([10]) introduced the use of the *blackboard architecture*. Multiple knowledge modules solving separate problem subtasks communicate with each other by posting messages on the blackboard (working memory). All knowledge modules have access to the blackboard and are given the opportunity to respond to posted messages. If more than one submodule is eligible to take action in response to the contents of the blackboard, a control module is engaged to determine what the system should do next. Hearsay was designed for speech recognition and was able to identify a vocabulary of more than 1000 words.

PROSPECTOR
An expert system used for mineral deposits exploration ([7]). PROSPECTOR represented causal knowledge using rules and taxonomic knowledge using frames. It used inference networks for propagating uncertainty measures.

INTERNIST/CADUCEUS
Performs diagnosis in the internal medicine domain ([21]). The system evaluates probabilities of the various diseases based on the symptoms and uses heuristics to inquire about additional symptoms. The most probable diseases are given as output.

FXAA
FXAA (Foreign Exchange Auditing Assistant) is an expert system used by Chemical Bank to detect irregularities in foreign exchange transactions. It was developed using the expert system shell ART.

GATES
GATES (Gate Assignment and Tracking Expert System) is used in the airline industry to design gate assignment plans for inbound and outbound flights. Gates was written in Prolog.

AUTHORIZER'S ASSISTANT
Used by American Express for approval of credit card charges. The system has reduced denial of credit decisions by one third resulting in

millions of dollars a year in additional profit for American Express.

SMART

SMART ([1]) is a problem resolution system used by Compaq in customer service requests. It is based on the *case based reasoning* paradigm. It contains sample cases of diagnostic problems on Compaq computers and when a new case is called in by a customer, the case closest to it is retrieved and used to generate a solution to the problem. Using the SMART system, the customer service department at Compaq was able to raise its rate of successful solution in one try of customer problems from 50% to 87%.

MOLE

MOLE, ([15]), is a second generation expert system shell that solves *diagnostic* problems by the *cover-and-differentiate* method. The cover-and-differentiate method finds possible causes explaining symptoms and then differentiates among the possible causes. It possesses a knowledge acquisition module. Obviously, for a problem to be a good fit to be solved by MOLE, the set of diagnostic problems or symptoms must be known and possible causes also known and there must be a way to differentiate among causes. The MOLE has been used among other applications to build expert systems for diagnosing problems in a steel rolling mill, and a coal burning power plant.

VT and SALT

SALT, ([16]), solves problems that involve *constraint satisfaction tasks*, like planning and configuration, by using the *propose-and-revise* method. It possesses a knowledge acquisition tool that performs knowledge elicitation and generates OPS5 type of rules. It can be used by the domain experts without the need for a knowledge engineer. SALT has been successfully used for scheduling manufacturing processes and VT is a SALT generated expert system used to design elevator systems at Westinghouse Elevator company.

ONCONCIN, OPAL

OPAL is a task-specific acquisition tool for the cancer chemotherapy domain, ([17]). It helps oncologists build cancer treatment plans that are used as a knowledge base in the expert system ONCOCIN. It requires its users to fill out slots in predefined forms using a predefined vocabulary of representation primitives.

SSCFI (Special Services Circuits Fault Isolation) and NETTRAC:

Both systems were developed by GTE. SSCFI performs problem diagnosis and service dispatching tasks in the telecommunications industry. Customers report problems with special service circuits. SSCFI determines which circuit tests to run, conducts the tests, diagnoses the problem and makes dispatch decisions. It is used in all of GTE's regional centers. It was implemented using the expert system shell ART-IM. NETTRAC is used for network traffic management. It uses network performance data, which it receives from switches, to recognize and interpret anomalies and plan and implement solutions.

KEFIR (KEy FIndings Reporter):
KEFIR is a generic automated knowledge acquisition tool also developed by GTE Labs. It was used to implement Health-Kefir, a system which reviews health care service use for GTE's 100,000 employees and their dependents. The system monitors and analyzes changes in the medical database. It identifies areas where costs are likely to increase next year, and where specific medical interventions could save the company money and improve the health of its employees.

PROSE:
Prose is a configuration system used to process and configure orders at AT&T. According to AT&T, it has dramatically reduced order-to-service intervals. It was built using the Classic Description Logic System.

EMCS (Expert Message Correction System):
EMCS investigates and recovers telecommunication charges from unidentified calls. It was built using ART-IM. Developed and used in Pacific Bell which reports the recovery of more than $1 million in charges annually.

REFERENCES

[1] Acorn, T., Walden, S., "SMART: Support Management Automated Reasoning Technology for Compaq customer service", *Proceedings of the 4th Innovative Applications of AI Conference*, 1992.

[2] Boose, J.H., Bradshaw,J.M., "Expertise transfer and Complex Problems: Using Aquinas as a knowledge acquisition workbench for Knowledge-Based Systems", *Int. J. Man-Machine Studies*, vol. 27, 1987, pp. 167-179.

[3] Boose, J.H., "A Knowledge Acquisition Program for Expert Systems Based on Personal Construct Psychology", *Int. J. Man-*

Machine Studies, vol. 23, 1985, pp. 495-525.

[4] Brownston, L., Farrell, R., Kant, E., Martin, N., *Programming Expert Systems in OPS5*, Addison Wesley, 1986.

[5] Davis, R., "Interactive Transfer of Expertise: Acquisition of new inference rules", *Artif. Intell.*, vol. 12, 1979, pp. 121-157.

[6] Dreyfus, H., Dreyfus, S., *Mind over Machine: The Power of Human Intuition and Expertise in the Era of the Computer*, New York, The Free Press, 1986.

[7] Duda, R., Gaschnig, J., Hart, P., "Model design in the Prospector Consultant system for mineral exploration", In: *Expert Systems in the Microelectronic age,* Ed., Michie, D., Edinburg Press, 1980.

[8] Durkin, J., "*Expert Systems: Catalog of Applications*", Intelligent Computer Systems, Inc., Akron, Ohio, 1993.

[9] Eriksson, H., "A survey of knowledge acquisition techniques and tools and their relationship to software engineering", *Journal of Systems and Software*, vol. 19, no 1, 1992, pp. 97-107.

[10] Erman, L., Hayes-Roth, F., Lesser, V., Reddy, D., "The Hearsay I Speech Understanding system: Integrating knowledge to resolve uncertainty", *ACM Computing Surveys*, 12, 2, 1980.

[11] Feigenbaum, E.A., "Themes and case studies of knowledge engineering", *Expert Systems in the Micro-Electronic Age*, Ed. D. Michie, Edinburg Univ. Press, 1979, pp. 3-25.

[12] Giarratano, J., Riley, G., *Expert Systems: Principles and Programming*, PWS-Kent, 1989.

[13] Holland, J.H., Holyoak, K.J., Nisbett, R.E., Thagard, P.R., *Induction: Processes of Inference, Learning and Discovery*, MIT Press, 1986.

[14] Lindsay, R., Buchanan, B., Feigenbaum, E., Lederberg, J., *Applications of Artificial Intelligence for Chemical Inference: The Dentral Project*, McGraw-Hill, NY, 1980.

[15] Marcus, S., *Automating Knowledge Acquisition for Expert*

Systems, Kluwer Academic, 1988.

[16] Marcus, S., McDermott, J., "SALT: A Knowledge Acquisition Language for propose-and-revise Systems", *Artif. Intell.*, vol 39, 1989, pp. 1-37.

[17] Musen, M.A., Fagan, L.M., Combs, D.M., Shortliffe, E.H., "Use of a domain model to drive an interactive knowledge editing tool", *Intern. J. of Man-Machine Studies*, vol. 26, #1, 1987, pp. 105-121.

[18] Newell, A., "The Knowledge Level", *Artificial Intelligence*, 18,1, 1982, 87-127.

[19] Newell, A., Simon, H.A., "GPS: A program that simulates Human Thought", *Computers and Thought*, ed. Feigenbaum, E., Feldman, J., NY, McGraw-Hill, 272-296.

[20] Oravec, J.O. and L. Travis, "If we could do it over, we'd...Learning from less-than-successful projects", *Journal of Systems and Software*, vol. 19, no 2, 1992, pp. 113-122.

[21] Pople, H., "CADUCEUS: An experimental expert system for medical diagnosis", In: Winston, P. and Prendergast, K. (eds.), *The AI Business*, MIT, Cambridge, MA, 1984, pp. 67-80.

[22] Shortliffe, E., *Computer Based Medical Consultations: MYCIN*, Elsevier, 1976.

[23] Simon, H., Newell, A., "Heuristic Problem Solving: The next advance in Operations Research", *Operations Research*, vol. 6, 1958, pp. 1-10.

[24] Stylianou, A., Madey, G., Smith, R., "Selection Criteria for Expert System Shells", *Communications of the ACM*, Vol. 35, No. 10, October 1992, pp. 28-48.

[25] Wielienga, B., Schreiber, G., Breuker, J., "KADS: A modeling approach to knowledge engineering", *Knowledge Acquisition*, 4,1, 1992.

CHAPTER 2

THEORETICAL FOUNDATIONS

2.1 INTRODUCTION

Expert Systems are a product of the symbolic approach to Artificial Intelligence and they are based to a large extent on ideas and principles from the theory of formal logic. An introduction to these foundational principles can contribute to a more thorough understanding and grounding on the subject of Expert Systems and the theoretical issues involved.

The birth of formal logic can be traced to ancient Greece and Aristotle, who introduced the notion that correct reasoning can only take place by the application of valid inference rules (called *syllogisms*). George Boole introduced Boolean Algebra in the 1800s, which is a propositional logic system based on the logical operators AND, OR, NOT. One of the most significant applications of Boolean Algebra was in the digital design of computer hardware. Later, Frege introduced predicate logic as a formalism to extend propositional logic through the use of quantifiers, functions and predicates ([6]).

While natural language can be ambiguous and context sensitive, knowledge expressed in the syntax of logic has a unique, well-defined meaning which when combined with a sound inferencing process enables correct reasoning. In scientific disciplines, the problem definition, the knowledge about the problem domain and the solution space, all need to be expressed accurately and concretely. Formal Logic provides such a vehicle for precisely expressing domain knowledge and a tool for manipulating knowledge through deductive reasoning techniques.

The realization of the representation and manipulation capabilities of logic in Computer Science, gave rise to the development of Logic Programming and Automated Theorem Proving. Kowalski, ([9]), suggested the procedural interpretation of declarative logic formulas for problem solving and Colmerauer and his research group at Marseilles implemented Kowalski's proposals into the programming language Prolog, (PROgramming with LOGic). A significant result along the way has been the introduction of the resolution principle (Robinson, [21]), which is used in a specialized form by the inference engine of Prolog.

In this chapter, only boolean and monotonic logic are discussed in adequate detail and attempts to introduce multi-valued and non-monotonic logic in expert systems are only briefly touched upon. Uncertainty, which renders the knowledge base non-monotonic, is covered in more detail in Chapter 6.

2.2 PROPOSITIONAL LOGIC

In propositional logic, logic formulas are constructed by combining true or false propositions with the basic logical operators of negation (¬), conjunction (∧), and disjunction (∨). A *proposition* is a statement having the value of either *true* or *false*. For example,

there_are_seven_days_in_a_week or *the_current_year_is_1795*

Other logical operators, like implication, (⇒), or logical equivalence, (≡), are reducible to the basic operators. For example, A⇒B is logically equivalent to ¬A∨B, and A≡B is equivalent to (A⇒B)∧(B⇒A) or (¬A∨B)∧(¬B∨A) or (¬A∧¬B)∨(A∧B).

Logic formulas can be proven by using truth tables, where all the possible combinations of the truth values of the variables are considered. The truth tables for the basic operators are given in Table 2.1.

Table 2.1 Truth table for basic logic operators.

p	q	p∨q	p∧q	¬p	p ⇒q
T	T	T	T	F	T
T	F	T	F	F	F
F	T	T	F	T	T
F	F	F	F	T	T

Propositional logic formulas can also be proven algebraically by using well known logical equivalences, such as:

$A \vee \neg A$ *is TRUE,* $A \wedge \neg A$ *is FALSE,*
$\neg(\neg A) = A$,
$A \wedge B = B \wedge A$, $A \vee B = B \vee A$ *(commutative property)*
$\neg(A \wedge B) = \neg A \vee \neg B$, $\neg(A \vee B) = \neg A \wedge \neg B$ *(deMorgan's Law)*
$A \vee (B \wedge C) = (A \vee B) \wedge (A \vee C)$, and
$A \wedge (B \vee C) = (A \wedge B) \vee (A \wedge C)$ *(distributive property)*
$A \wedge (B \wedge C) = (A \wedge B) \wedge C$, $A \vee (B \vee C) = (A \vee B) \vee C$ *(associative property)*
$A \Rightarrow B = \neg A \vee B$

As early as 1956, Newell, Shaw and Simon, [14], built an automated proof system for propositional logic. The Logic Theorist, as they called it, was able to prove many of the theorems in Whitehead and Russell's Principia Mathematica. Logic Theorist assumed five axioms and used the following techniques to prove a propositional logic formula:

a) To show P, find an axiom or theorem of the form A⇒P and try to show A,
b) To show A⇒P, find an axiom or theorem of the form A⇒B and show B⇒P (*forward chaining*),
c) To show A⇒P, find an axiom or theorem of the form B⇒P and show A⇒B (*backward chaining*).

The topic of propositional logic is not expanded further, as we assume the reader's familiarity with it.

2.3 FIRST ORDER PREDICATE CALCULUS AND PREDICATE LOGIC

2.3.1 First Order Predicate Calculus

The expressive power of propositional logic is limited. Propositional logic is unable to express the concept of an individual object, an object property or relations between objects. For example, the inference chain {"all humans are mortal", "there is at least one human", therefore "there is at least one mortal"} cannot be efficiently expressed in propositional calculus format. Boolean variables would have to be used to describe whole sentences, much of the knowledge would be left implicit and the meaning of the quantifiers *"all"* and *"there is"* would be lost in the propositional representation.

First order predicate calculus extends the expressive power of propositional logic and can be used as a formal language to express declarative knowledge. A predicate calculus formula consists of *predicates*, predicate *arguments*, *logical operators* and *quantifiers*. A *predicate* is a relation that maps objects/arguments in the universe of discourse into TRUE or FALSE. For example: *father(chris,basil)* is true, if *chris* is the father of *basil*. "*father*" can be defined as a predicate of two arguments, which returns true if the first argument is the father of the second argument. The *arity* of a predicate is its number of arguments. In predicate calculus, arguments can be constants representing objects from the domain of discourse or variables. From now on, we will use a character string with its first letter in lower case to denote a constant, while an upper case first letter indicates a variable.

An *atomic formula* is a predicate of a number of arguments. Predicates are joined together to form logical sentences using the *logical connectives*

and ($\wedge$), *or* ($\vee$), *not* ($\neg$), *implies* ($\Rightarrow$)

and *quantifiers*, which qualify the variable predicate arguments. The *existential quantifier* is denoted by $\exists$, and the *universal quantifier* by $\forall$.

The logic formula: $\forall X\ p(X)$ has the semantic interpretation that for every object X in the domain of discourse, X has the property p or makes the logical proposition p true. For example, "all birds can fly" can be expressed in predicate calculus as:

$\forall X\ (bird(X) \Rightarrow fly(X))$.

The formula: $\exists X\ p(X)$ means that there is an object X, (at least one), in the universe of discourse which satisfies the property p. For example, *"there is an odd, prime number"* can be expressed as:

$\exists X\ (number(X) \wedge prime(X) \wedge odd(X))$.

We give below some examples where knowledge expressed in natural language is converted to predicate calculus formulas.

The phrase "*every human is mortal*" can be expressed in predicate calculus as:

$\forall X (human(X) \Rightarrow mortal(X))$,

where "human" is a predicate of one argument X, which becomes true if variable X is assigned an object from the domain of discourse which is a human and "mortal" is a predicate of one argument X, which becomes true if X is mortal.

The phrase "*every person has exactly two parents*" can be expressed in predicate calculus format as:

$\forall P(\exists X \exists Y\ (parent(X,P) \wedge parent(Y,P) \wedge \neg equal(X,Y) \wedge (\forall Z parent(Z,P) \Rightarrow [equal(Z,X)\ \vee equal(Z,Y)])))$

As another example, consider the phrase "*every man loves a woman*". It can be written in predicate calculus as:

$\forall X\ (man(X) \Rightarrow \exists Y\ (woman(Y) \wedge loves(X,Y)))$.

The phrase "*there is a red pyramid on top of every cube*" can be written as:

$\forall X\ (cube(X) \Rightarrow \exists Y\ (pyramid(Y) \wedge color(Y,red) \wedge ontop(Y,X)))$.

As another example, consider "*there is no car as good as a mercedes*". It can be written as:

$\neg\exists X\ [car\_model(X) \wedge \neg equal(X,mercedes) \wedge asgoodas(X,mercedes)]$

or equivalently

$\forall X\ ((car\_model(X) \wedge \neg equal(X,mercedes)) \Rightarrow \neg asgoodas(X,mercedes))$.

2.3.2 First order predicate logic, functions

First order predicate logic extends predicate calculus by introducing functions.

A *function* is a relation which maps objects from the domain of discourse to other objects and it can take the place of an argument of a predicate. For example, consider the predicate *father(X,Y)* of two arguments which assumes the value *true* when X is the father of Y and the predicate *husband(X,Y)* which becomes *true* when X is the husband of Y. Then the logic formula

$\forall X\ \forall Y\ (father(X,Y) \Rightarrow husband(X,mother(Y))$

states that "if X is the father of Y, then X is the husband of the mother of Y", where the function *mother* accepts as input the name of an individual and returns the name of his/her mother.

As another example, consider a set of blocks lying on a table. In order to describe the fact that if block A is put on top of block B, then block A is on top of B, the expression:

ontop(A,B, puton(A, B, S))

can be used, where *ontop* is a predicate of arity three and *puton* a function of three arguments. The first two arguments of the predicate *ontop* are blocks, and the third is a state describing a certain configuration of blocks on the table. The predicate *ontop(A, B, S)* is true if block A is on top of block B in the block configuration described by state S. The function *puton* accepts as input two blocks and a current state S. It produces as output the new state resulting from the current state S after block A has been placed on top of block B. The formula *ontop(A,B, puton(A,B,S))* might be attached the semantic meaning of: "given a state S, if we put block A on top of block B then in the resulting state *puton(A,B,S)*, block A will be on block B.

First order predicate logic extends predicate calculus by introducing functions. Functions can also be used as predicate arguments. First order predicate logic does not allow predicate or function names to be variables. This makes statements like "there is an object which satisfies all properties in the domain" difficult to express in first order predicate logic, (we would have to write a sentence for each possible predicate in the domain). In second order predicate calculus, this restriction is lifted and predicates or functions can be variables themselves. Second order predicate logic is beyond the aims of this book and is not covered in this chapter.

In the next section, we describe what constitutes a syntactically correct first order predicate logic formula and define the scope of quantifiers.

2.3.3 Well formed formulas

An atomic formula or its negation is called a *literal*. Literals are joined by the logical operators and quantifiers to form *well formed formulas*, *(WFF)*. Informally, a *well-formed formula* is recursively defined as follows:

1. A literal is a WFF
2. If *g* and *h* are two WFF, then *g<op>h* is also a WFF, where <op> is a logical operator.
3. If *g* is a WFF, then $\forall$X (*g*) and $\exists$X(*g*) is a WFF.

The syntax of a *well-formed formula* can be defined more formally by the grammar (G,N,T,R), where N is the set of nonterminals (denoted by <name>), T is the set of terminals and R is the set of production rules. Terminal symbols include the quantifiers, logical operators, a set of predicate names P, a set of variables V, a set of constants C and a set of function names F.

The rule set R is given by:
<atom> → <predicate>(<arguments>)
<literal> → ¬ <atom> / <atom>
<quantifier> → ∀ / ∃
<operator> → ∧ / ∨ / ⇒
<predicate> → p, for each p in the predicate set P
<variable> → X, for each X in the variable set V
<constant> → c, for each object c in the universe of discourse C
<function_name> → f, for each f in the set of functions F
<function_term> → <function_name> (<arguments>)
<argument> → <variable> / <constant> / <function_term>
<arguments> → <argument>,<arguments> / <argument>
<sentence> → (<literal><operator><literal>)
/ <quantifier><variable><literal> / <literal>

Given a quantified formula $\forall X(p)$, all occurrences of the variable X in *p* are said to be in the *scope* of the universal quantifier. A variable not in the scope of any quantifier is a *free* variable. A quantified variable is bound to its closest quantifier. For example, in the formula

$$\forall X\ (p(X,Y) \Rightarrow \exists Z q(Z))$$

the variable X in p(X,Y) is bound to the universally quantified variable

X, Z is bound to existential quantified Z and Y is a free variable. If a variable is in the scope of more than one quantifier, then it is bound by the innermost quantifier.

For example, in the formula

$$\forall X\ (p(X,Y) \Rightarrow \exists X q(X))$$

the variable X in q(X) is in the scope of both $\forall$ and $\exists$, but it is bound to the existential quantifier. The second variable X is different than the first and by renaming it Z, the formula

$$\forall X\ (p(X,Y) \Rightarrow \exists Z q(Z))$$

is equivalent to the one above.

When interpreting a well-defined formula which is not fully parenthesized, the following order of precedence is followed (from high precedence to low):

$\neg, \forall, \exists$ not, for every, there is
$\wedge$ and
$\vee$ or
$\Rightarrow$ implies

For example, the formula

$$\forall X\ (\neg p(X) \wedge \neg\exists Y\ q(X,Y) \vee s(X))$$

corresponds to the fully parenthesized formula:

$$\forall X\ (((\neg p(X)) \wedge \neg(\exists Y\ q(X,Y)) \vee s(X)).$$

2.4 INFERENCE

Inference is the process of deriving conclusions from a set of premises. Each inference step in the deduction chain must be based on a logically valid *rule of inference.*

The inferencing process starts with a given set of well formed logical formulas, called the *knowledge base* or the *set of axioms*, and in each inferencing step the current state of the knowledge base is transformed into a new state with additional WFFs inferred by using a "rule of inference". In essence, inference can be viewed as a state-space problem. The set of well-formed formulas defines a state, and the rules of inference correspond to operators which transfer the current state to the next state by appending new valid formulas to the knowledge base. The goal state is a state which contains the theorem to be proven among its WFFs.

Here are some examples of valid inference rules.

Modus ponens:

Given the premises A⇒B and A, then it can be inferred that B is true.

A⇒B
A

———

.: B

Modus tollens:

Given the premises A⇒B and ¬B, (B is false), then it can be inferred that ¬A, (A is false).

A⇒B
¬B

———

.: ¬A

Resolution:

Given the premises A∨B and ¬B∨C, then it can be inferred that A∨C is true.

A∨B
¬B∨C

———

.: A∨C

All of the deductive rules given above can be easily proven by constructing the truth tables or by using the distributive properties of the logical *and*, *or* and the equivalent of deMorgan's laws for boolean logic.

Observe that both modus ponens and modus tollens are special cases of resolution, as A⇒B is logically equivalent to ¬A∨B. For example, if A⇒B and A are true, then ¬A∨B and A are true. Applying resolution, we get that B is also true.

2.5 PROOF BY REFUTATION RESOLUTION

The inference strategy used in Prolog and in the inference engine of numerous expert systems is a restricted form of resolution. Resolution is an inference method which is both sound and complete. The inference mechanism used in a reasoning system is *sound* if and only if it is logically valid, i.e., valid premises give rise to a valid conclusion. An inference rule is *complete* if and only if for each theorem which can be deduced from the knowledge base, the inference mechanism is able to deduce it from the premises. Modus ponens and modus tollens are sound, but they are not complete. For example, given that $a \Rightarrow b$ and $\neg b$, modus ponens cannot deduce that $\neg a$ is a valid conclusion.

A *rule base* is composed of axioms, (facts and rules). A *rule* is a well-formed formula of the form $A \Rightarrow B$, where A and B are themselves well-formed formulas. A *fact* can be considered as a rule whose antecedent is always true. In reasoning systems, a query is posed to a knowledge base and the inference mechanism takes over, trying to deduce (prove) the query. In logic programming, the query is usually called a *theorem*. A "theorem" is proven from the "axioms" using valid logical deductive steps.

In a resolution based proof system, before a query can be answered, all rules must be converted into "clause" form. A logic formula is in *clause* form if it is a disjunction of literals.

In what follows, we first present an algorithm for converting a logic formula into clause form. We then introduce the unification process which prepares clauses for the application of resolution, followed by an explanation of the resolution based, proof by refutation method.

2.5.1 Conversion of axioms into clause form

The following formulas are easily verified and are used in the conversion process. Let A and B be first order predicate logic expressions. Then:

$$A \Rightarrow B \;\; iff \;\; \neg A \vee B \qquad (2.1)$$
$$\neg(A \wedge B) \;\; iff \;\; \neg A \vee \neg B \qquad (2.2)$$
$$\neg(A \vee B) \;\; iff \;\; \neg A \wedge \neg B \qquad (2.3)$$
$$\neg(\forall X\, p(X)) \;\; iff \;\; \exists X(\neg p(X)) \qquad (2.4)$$
$$\neg(\exists X\, p(X)) \;\; iff \;\; \forall X(\neg p(X)) \qquad (2.5)$$

$\neg A \wedge A$ *is false* (2.6)
$\exists X\ p(X)$ *iff* $p(c)$, *for some constant c* (2.7)
$\forall X\ \forall Y\ \exists Z\ p(X,Y,Z)$ *iff* $\forall X\ \forall Y\ p(X,Y,g(X,Y))$ (2.8)
$A \vee (B \wedge C)$ *iff* $(A \vee B) \wedge (A \vee C)$ (2.9)

In (2.8), g is a function which given X and Y, calculates the value of Z for which p(X,Y,Z) becomes true. The function *g* is called a *Skolem function.*

To ***convert an axiom to clause form*** *do the following*:

1. Eliminate implications by using Proposition (2.1).
2. Distribute negations so that they apply to single atomic formulas by using propositions (2.2), (2.3), (2.4), (2.5).
3. Rename variables if necessary, so that a variable is not quantified more than once. (if a variable is in the scope of two or more quantifiers, then it is bound to the innermost quantifier)
4. Purge existential quantifiers by using skolemization, as in (2.7), (2.8).
5. Purge universal quantifiers (since all variables are universally quantified, the quantifiers can be omitted).
6. Write the formula as a conjunction of disjunctions of literals using Proposition (2.9).
7. Eliminate conjunctions by substituting a conjunction by its individual conjuncts.
8. Rename variables if necessary, so that the same variable does not appear in more than one clause.

The steps involved in the procedure above can be better explained through the use of an example.

Consider the following description in natural language:

If AIONICS employees do not belong to any union, then it is not true that every employee of AIONICS has to pay a certain amount for union dues.

Consider the predicates:
e(X) --- X is an employee of AIONICS
b(X,Y)---employee X is a member of Y
p(X,Z)---employee X pays amount Z for union dues
u(X)---- X is a union

then the given english statement can be expressed in predicate calculus as follows:

$$(\forall X\ \forall Y [e(X) \wedge u(Y) \Rightarrow \neg b(X,Y)]) \Rightarrow [\neg \forall X\ (e(X) \Rightarrow \exists Z\ (p(X,Z))] \qquad (2.10)$$

Conversion to clause form:

Step 1: Eliminate the implications.
To perform this step use the fact that A⇒B is logically equivalent to ¬A∨B. After eliminating the third implication, the axiom becomes:

$$(\forall X\ \forall Y\ [e(X) \wedge u(Y) \Rightarrow \neg b(X,Y)]) \Rightarrow [\neg \forall X\ (\neg e(X) \vee \exists Z\ (p(X,Z))]$$

After eliminating the first implication the axiom becomes:

$$(\forall X\ \forall Y\ [\neg(e(X) \wedge u(Y)) \vee \neg b(X,Y)]) \Rightarrow [\neg \forall X\ (\neg e(X) \vee \exists Z\ (p(X,Z))]$$

After eliminating the remaining implication, the axiom becomes:

$$\neg(\forall X\ \forall Y\ [\neg(e(X) \wedge u(Y)) \vee \neg b(X,Y)]) \vee [\neg \forall X\ (\neg e(X) \vee \exists Z\ (p(X,Z))]$$

Step 2: Move negations so that they negate atomic formulas.
The axiom becomes:

$$\exists X\ \exists Y\ \neg(\neg e(X) \vee \neg u(Y) \vee \neg b(X,Y)) \vee [\exists X\ \neg(\neg e(X) \vee \exists Z\ (p(X,Z))]$$
or
$$\exists X\ \exists Y\ (e(X) \wedge u(Y) \wedge b(X,Y)) \vee [\exists X\ (e(X) \wedge \neg \exists Z\ (p(X,Z))]$$
or
$$\exists X\ \exists Y\ (e(X) \wedge u(Y) \wedge b(X,Y)) \vee [\exists X\ (e(X) \wedge \forall Z\ (\neg p(X,Z))]$$

Step 3: Rename variables.
While the variable X in the first term is universally quantified, the variable X in the second term is existentially quantified and clearly different from X. To avoid confusion, rename the existentially quantified X as W to get:

$$\exists X\ \exists Y\ (e(X) \wedge u(Y) \wedge b(X,Y)) \vee [\exists W\ \forall Z\ (e(W) \wedge (\neg p(W,Z))]$$

Step 4: Eliminate existential quantifiers.
This is done by introducing Skolem functions, ([16]), based on the observation that

$$\forall X\ \exists Z\ p(X,Z) \quad iff \quad \forall X\ p(X,f(X))$$

where f(X) is a function which maps X to the necessary Z to make the property p(X,Z) true, i.e., f(X)=Z.

The value of Z in the above formula depends on the value of Z, and Z is called a variable "*bound*" by X, (or in the "scope" of X).
In the formula:

$$\exists Z\ p(Z)$$

the variable Z is not dependent on another variable, so the skolemization of this formula is

$$p(c)$$

where c is some constant.

In the employee formula, skolemization gives the following formula:

$$[e(c1\ ) \wedge u(c2) \wedge b(c1,c2)] \vee [\forall Z\ (e(c3) \wedge \neg p(c3,Z))]$$

where c1,c2,c3 is the skolemization of X,Y and W.

Step 5: Move universal quantifiers to the left and purge them.
Since, after Skolemization, all remaining variables are universally quantified, and the variables were renamed, all the universal quantifiers can now be moved to the left. We thus get the following equivalent formula:

$$\forall Z\ [e(c1) \wedge u(c2) \wedge b(c1,c2)] \vee [(e(c3) \wedge \neg p(c3,Z))]$$

Step 6: Move disjunctions down to literals.
In this step the distributive property is used, $A\vee(B\wedge C)$ iff $(A\vee B)\wedge(B\vee C)$.
The employee formula now becomes:

$\forall Z\ [e(c1) \vee \neg p(c3,Z)] \wedge [e(c1) \vee e(c3)] \wedge [u(c2) \vee \neg p(c3,Z)]$
$\wedge [u(c2) \vee e(c3)] \wedge [b(c1,c2) \vee \neg p(c3,Z)] \wedge [e(c3) \vee b(c1,c2)]$

Step 7: Rewrite conjunctions as separate clauses, since A∧B iff A is true, B is true. We get the clauses:

$[e(c1) \vee \neg p(c3,Z)]$
$[e(c1) \vee e(c3)]$
$[u(c2) \vee \neg p(c3,Z)]$
$[u(c2) \vee e(c3)]$
$[b(c1,c2) \vee \neg p(c3,Z)]$
$[e(c3) \vee b(c1,c2)]$

Step 8 and 9: Rename the variables to finally get a knowledge base of clauses which is equivalent to the employee formula:

$e(c1) \vee \neg p(c3,Z)$
$e(c1) \vee e(c3)$
$u(c2) \vee \neg p(c3,W)$
$u(c2) \vee e(c3)$
$b(c1,c2) \vee \neg p(c3,X)$
$e(c3) \vee b(c1,c2)$

2.5.2 Unification

In order to prepare two clauses for resolution, the clauses may need to be unified. Two literals P and Q are called *unifiable* iff argument variables can be substituted by constants or other variables in such a way that the two literals P, Q become identical.

For example, given the literals

alumnus(X, bradley, Y) and *alumnus(hersey, Z, W)*

the substitution of X by hersey, Z by bradley, Y by 1989 and W by 1989, (X←hersey, Z←bradley, Y←1989, W←1989), will unify the two literals into the literal:

alumnus(hersey, bradley, 1989).

The substitution which unifies two literals is called a *unifier.* A unifier is not necessarily unique. For example, the substitution (X←hersey, Z←bradley, Y←W) also unifies the literals above into the literal

alumnus(hersey, bradley, W).

This literal is more general than the one derived above. Literal A is said to be *more general* than literal B if there is at least one substitution which reduces A to B. A literal L is said to be *a most general literal* iff there is no substitution *s* and no other literal M such that *s* reduces M to L.

2.5.2.1 Finding the most general unifier

A *most general unifier* for literals P and Q is a substitution which unifies the two literals into a most general literal. Equivalently, a most general unifier substitution for literals P and Q is one for which no subset of substitutions also unifies the literals.

The most general unifier of two literals is not necessarily unique. A well known procedure for finding a most general unifier for a set C of literals is given below.

1. Let k=0, S={∅}
(eventually S holds the most general unifier substitution for C).

2. If the set of substitutions in S applied to C gives only one literal,
then
S is the most general unifier of C.
else
find the first position of each literal where corresponding symbols differ.

3. If the differing symbols in that position are both variables,
then
exchange the symbols, let S=S∪{new substitution} and go to 2
else
If one is a variable and the other is a constant or function which does not contain the variable
then
substitute the variable by the constant or function
else

If both are constants
then report that clauses are not unifiable.

2.5.2.2 Unification examples

As an example consider the clauses:

p(f(X),Y,g(Y)) and *p(f(X),Z,g(X))*

A linear scanning of the clauses left to right will produce the substitutions:

Y←Z: p(f(X),Z,g(Z)) and p(f(X),Z,g(X)),

followed by Z←X: p(f(X),X,g(X))

The most general unifier of the clauses is: p(f(X),X,g(X)) .

As another example consider the clauses:

p(f(X),Y) and p(Y,Z).

Applying the unification algorithm yields:

Y←f(X): p(f(X),f(X)) and p(f(X),Z)
Z←f(X): p(f(X),f(X)) which is the most general unifier.

2.5.3 Resolution proofs by refutation

After all the axioms in a knowledge base have been converted into clause form, inference can take place using resolution. To prove a theorem, (query), using resolution do the following:

1. *Add the negation of the theorem to the set of clauses in the knowledge base.*
2. *Search for two resolvable clauses among the set of axioms.*
   If no pair of resolvable clauses can be found,
   then
   exit and report that the theorem to be proven is false,
   else

resolve the clauses and add the resolvant clause into the set of axioms.

3. *If a contradiction is reached, then the theorem is true, else go to 2.*

The following example demonstrates the theorem proving technique described above.

Example:
Hersey Eagles is a basketball player for the Seattle Supersonics and does not attend or teach any classes at Bradley University. Hersey is allowed to get Bradley Braves basketball tickets at a discount. At a university, only students, faculty and alumni can purchase tickets for university basketball games at a discount. Hersey is an alumnus of Kim's tae-kwon-do academy. Show that Hersey is a Bradley University alumnus.

Consider the following predicates:

playbasketball(X,Y) is true if X plays basketball in team Y.
student(X,Y) is true if X is a student at university Y.
discount(X,Y) is true if X can get tickets at discount at Y.
faculty(X,Y) is true if X is a faculty member at Y.
alumnus(X,U) is true if X is an alumnus of U.
university(X) is true if X is a university.

The predicate calculus knowledge base describing the above situation is given by (irrelevant knowledge like who is the tae-kwon-do master was omitted):

discount(hersey,bradley).
¬faculty(hersey,bradley).
playbasketball(hersey,supersonics).
¬student(hersey,bradley).
university(bradley).
alumnus(hersey,tae-kwon-do).
$\forall X\ \forall U\ [university(U) \wedge (student(X,U) \vee faculty(X,U) \vee alumnus(X,U))]$
$\quad \Rightarrow discount(X,U).$
$\forall X\ \forall U\ [university(U) \wedge \neg(student(X,U) \vee faculty(X,U) \vee alumnus(X,U))]$
$\quad \Rightarrow \neg discount(X,U).$

To this set of axioms, we append the negated query:

¬alumnus(hersey,bradley).

Before we apply resolution, we convert the axioms in the knowledge base into clause form, which yields the following clauses:

$$playbasketball(hersey, supersonics). \tag{2.11}$$
$$discount(hersey, bradley). \tag{2.12}$$
$$\neg faculty(hersey, bradley). \tag{2.13}$$
$$\neg student(hersey, bradley). \tag{2.14}$$
$$university(bradley). \tag{2.15}$$
$$\neg university(U) \vee \neg student(X, U) \vee discount(X, U). \tag{2.16}$$
$$\neg university(U) \vee \neg faculty(X, U) \vee discount(X, U). \tag{2.17}$$
$$\neg university(U) \vee \neg alumnus(X, U) \vee discount(X, U). \tag{2.18}$$
$$\neg university(U) \vee student(X, U) \vee faculty(X, U) \vee alumnus(X, U) \vee \neg discount(X, U). \tag{2.19}$$
$$\neg alumnus(hersey, bradley). \tag{2.20}$$

In order to resolve two clauses, one of them must contain the negation of a literal which also appears in the other clause.

For example, since all the variables are universally quantified, we can substitute the variable U in clause (2.19) by the constant "bradley" and the variable X by "hersey" to get the new, less general clause :

$$\neg university(bradley) \vee student(hersey, bradley) \vee faculty(hersey, bradley) \vee alumnus(hersey, bradley) \vee \neg discount(hersey, bradley) \tag{2.21}$$

Now clauses (2.21) and (2.20) are resolvable, and their resolution gives the new clause:

$$\neg university(bradley) \vee student(hersey, bradley) \vee \vee faculty(hersey, bradley) \vee \neg discount(hersey, bradley) \tag{2.22}$$

Resolving clause (2.22) with (2.12) gives the clause :

$$\neg university(bradley) \vee student(hersey, bradley) \vee faculty(hersey, bradley) \tag{2.23}$$

Resolving (2.23) with (2.15) gives the clause:

$$student(hersey, bradley) \vee faculty(hersey, bradley) \tag{2.24}$$

which in turn resolved with clause (2.14) gives the clause:

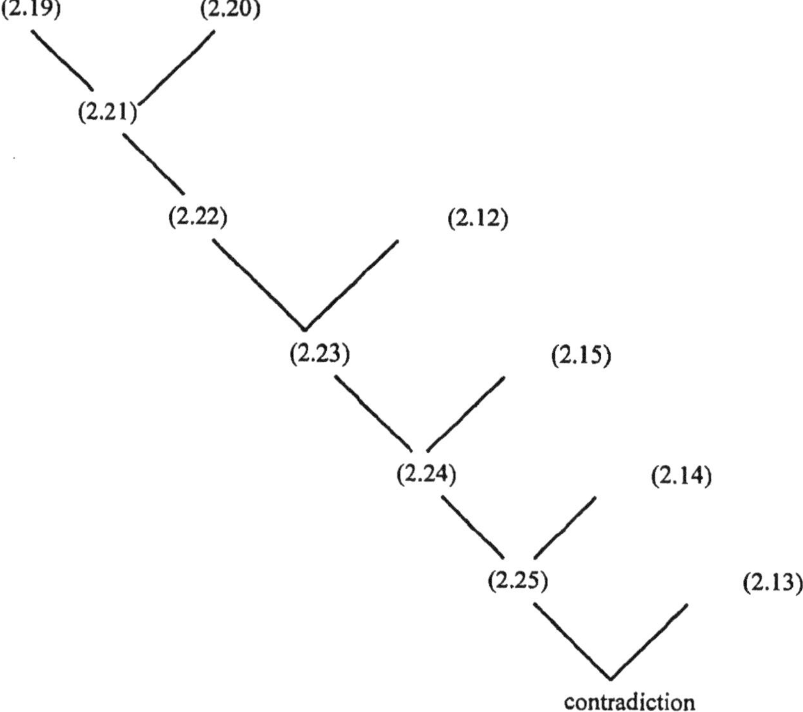

Figure 2.1 Resolution tree for the Hersey Eagle knowledge base.

faculty(hersey, bradley) (2.25)

Finally, clause (2.25) resolved with (2.13) gives the sought after contradiction. Therefore, since the negation of the query renders the knowledge base inconsistent, the query must be true. Such a resolution proof is commonly depicted by a resolution tree, as shown in Figure 2.1.

2.6 GREEN'S ANSWER TERM

Green introduced a method to retrieve the unification bindings that took place during the proof of a theorem, in effect giving the theorem proving system a query answering capability, ([7]). This is accomplished by appending an extra disjunction to the negated theorem, *the answer term*, containing all the variables whose values we want to retrieve.

For example, in the Hersey Eagle knowledge base, suppose that instead of just wanting to know if Hersey was a Bradley alumnus, we wanted to find all the Bradley alumni. In that case the query: *alumnus(X,bradley)* would be issued. Green's method appends to the knowledge base the disjunction of the negated query with an answer term. The answer term traps the bindings of the query variables. The resolution proof is not terminated when an empty clause is reached, but instead it is terminated when a clause containing the answer term as its only literal is reached.

In our example, the negated query with the answer term attached is:

$\neg alumnus(X, bradley) \lor answer(X)$

This clause resolves with (2.19) (after the unification substitution U=bradley in 2.19) to give:

$\neg university(bradley) \lor student(X, bradley) \lor faculty(X, bradley)$
$\lor \neg discount(X, bradley) \lor answer(X).$

Unifying this newly derived clause with (2.17) using the substitution U=bradley in (2.17) and resolving, we get the clause:

$\neg university(bradley) \lor student(X, bradley) \lor answer(X)$

which resolved with (2.15) gives:

$student(X, bradley) \lor answer(X)$

Finally, unify the new clause with (2.14) by using the substitution X=hersey to get the answer clause by itself:

$answer(hersey).$

The constants in the answer term contain the answers to the query "who are the alumni of Bradley University?".

2.7 KNOWLEDGE ACQUISITION BOTTLENECK

One of the major bottlenecks in expert system development is the knowledge acquisition bottleneck. For certain domains it may be from extremely difficult to almost impossible to build an accurate knowledge base which models the real world domain of expertise. Even in the simplistic example given above, the fact that Bradley is a university is so obvious and embedded in the human expert's conscience that it is very possible that the knowledge engineer may not include it in the knowledge base. Without the predicate "university" though, the database would not be complete and accurately depicting the situation. The query alumnus(hersey,bradley) would not have been provable. The topic of knowledge acquisition in expert systems is covered in detail in Chapters 4 and 5.

Machine learning is an area of Artificial Intelligence which has as its goal to automate knowledge acquisition, bypassing the knowledge acquisition bottleneck. In Chapter 7, we present an introduction to machine learning algorithms.

2.8 SEARCH STRATEGIES

Another major problem, which hinders the effectiveness of logic programming systems, is finding the appropriate search strategy to shorten the proof of a theorem. An exhaustive (brute force) search can be prohibitive in both time and space considerations. A search strategy needs to be employed which determines the clauses that are to be chosen for resolution, among the many possible resolvable pairs of clauses. A search strategy can make a very important contribution to the overall effectiveness of the system. A good search strategy significantly shortens the time needed to reach a proof, by not involving resolutions of clauses irrelevant to the proof. We give below some examples of search strategies which have been proposed in the literature:

Set of support
Let S be the set of clauses. Partition S into two sets K and S-K, where K is called the *support set*. Usually, K is chosen as the negated query. The set of support strategy requires that at least one of the clauses involved in each resolution step must be a member of K. Every time a new clause, C, is generated as a product of a resolution, the set of support set K is expanded to include the new clause C.

Linear resolution
One of the resolvents is either a clause in the original knowledge base or an ancestor of the other resolvent in the resolution tree.

SL-resolution
A procedure describing SL-resolution is given below:

Let C be a set of clauses and p the theorem to be proven.
While no contradiction is reached do
begin
Find a clause of the form p∨x1∨x2∨...∨xn call it current and resolve it with ¬p to get x1∨x2∨...∨xn. Let current=x1∨x2∨...∨xn.
While more literals in current and possible to delete do
begin
delete the first literal xi in current by resolving it with a clause of the form ¬xi∨y1∨y2∨...∨ym. Update current to be equal to the resolvant clause.
end
end

Breadth first search
Resolve all possible pairs of initial clauses, then all pairs of the new set of derived and old clauses, and keep repeating until a contradiction is reached.

Depth first search
Search the proof tree in depth first manner.

Unit preference strategy
Resolve clauses with fewer number of literals first.

Combinations of strategies are also possible. For example, SL-resolution is a combination of linear resolution and set of support.

All of the strategies described above are complete, (they are guaranteed to find a proof, if one exists), but they are subject to combinatorial explosion, (as the number of clauses increases linearly, the number of resolutions during a search for a proof increases exponentially), and the halting problem, (if there is no proof, the strategy may not be able to detect it and halt).

For a more detailed exposition of the strategies described above, as well as others, see Wos, ([23]), Loveland, ([10]), Chang and Lee, ([3]), Smith and Genesereth, ([22]). To solve the combinatorial explosion problem, various heuristics have been introduced which guide the search for a proof. Other researchers have addressed the search problem in logic programming (Kowalski, [8], Minker, [16], Nikolopoulos, [15]). In Kowalski's Σ^* algorithm, the order in which the clauses are resolved is determined by an evaluation function, which depends on the length of a clause and its level in the search tree. In Nikolopoulos's C* algorithm, a semantic measure of clause similarity is introduced based on entropy and is used to define a distance metric between clauses. The measure is then used in the context of logic programs to precluster the clause set and to determine the order in which they participate in the resolution process.

2.9 NON-MONOTONIC REASONING

The first order predicate logic model has certain drawbacks when applied to problem solving in the real world. The model assumes that a knowledge base is complete, consistent and certain, something that often is not the case. Real world knowledge usually is not boolean in nature. Even so, humans have no difficulty reasoning in the presence of uncertainty, vagueness and incompleteness. Several approaches exist for handling uncertain reasoning in expert systems, with the most notable being the Bayesian, the Dempster/Schaffer, Belief Networks and the fuzzy logic approach. A treatment of uncertainty in expert systems will be presented in Chapter 6. Another weakness of first order predicate logic and deductive reasoning is that knowledge bases of rules and inference techniques will never produce new knowledge about the world, they just convert implicit knowledge into explicit knowledge using deduction. Humans often use analogical and inductive reasoning to acquire new knowledge. In analogical reasoning a problem is solved by adopting a solution from a similar problem, which is already solved and inductive reasoning deals with the derivation of generalized knowledge about the problem domain from a set of instances or observations. Machine learning is the area of A.I. addressing the issue of learning new domain knowledge from a data set of instances and will be discussed in Chapters 7 and 8.

In an inferencing system which is based on predicate logic, if E is a knowledge base derived from D by using inference rules or by adding a new fact or rule then D is a subset of E. Furthermore, the completion of D is a subset of the completion of E, where the completion of a knowledge base D is defined as the union of D with all the clauses deducible from D. We express this by saying that the predicate logic formalism is a *monotonic* model of logic. On the other hand, conclusions in human reasoning are tentative and can be withdrawn after acquisition of new information. As a result, human logic is *non-monotonic* since the enlargement of the knowledge base does not necessarily enlarge its completion.

For example, knowing that "tweety is a bird" and that "birds can fly", one concludes that "tweety can fly". Upon learning that tweety happens to be a penguin and knowing that "penguins cannot fly", the conclusion can be reached that tweety cannot fly after all. Several things took place in the above reasoning process, which cannot be handled by monotonic logic. For example, in the default type of reasoning as in the one above, conclusions have a plausibility associated to them. Using the *default rule* "birds can fly", a human concludes that tweety can fly, but remains aware of the plausibility of the conclusion, in the sense that it may be overturned if more information becomes available. In contrast, deductive logic based reasoning assumes that the knowledge base is complete and consistent and a conclusion always remains true once it has been reached.

Simulating the simple reasoning process given above in a resolution based, predicate logic deductive system is difficult. For example, the knowledge base:

(1) $bird(X) \Rightarrow fly(X)$
(2) $bird(X) \wedge penguin(X) \Rightarrow \neg fly(X)$
(3) $bird(tweety)$
(4) $penguin(tweety)$

is both incomplete and inconsistent. Both of the conclusions $fly(tweety)$ and $\neg fly(tweety)$ can be reached. An attempt to fix it by replacing the first rule by $\text{bird}(X) \wedge \neg\text{penguin}(X) \Rightarrow \text{fly}(X)$ destroys the default type of reasoning, since in that case the default rule is lost and knowing that something is a bird no longer implies that it flies.

Non-monotonic Logic models incomplete knowledge bases and reasoning with incomplete knowledge (McCarthy, [11], Reiter, [18], McDermott and Doyle, [12]). It allows conclusions and all derivations involving those conclusions to be retracted in case new knowledge becomes available which necessitates reevaluation of beliefs. Methods to handle non-monotonicity of logic include using *default rules* and the employment of the *closed world assumption*. In the *closed world assumption* approach, (CWA), an atomic formula which is not deducible from the knowledge base is assumed to be false. In the next section, we will expand on the CWA approach.

In *default reasoning*, default rules are of the form:

a(X):b(X)/c(X).

A default rule is interpreted as follows: If $a(x_0)$ is deducible for some x_0 and $b(x_0)$ is not inconsistent, then conclude $c(x_0)$.

For example given the default rule

bird(X):¬penguin(X)/fly(X)

and the fact *bird(tweety)*, then *fly(tweety)* will be implied since *¬penguin(tweety)* is not inconsistent with the current knowledge. On the other hand, given the same default rule and the facts *bird(tweety)* and *penguin(tweety)*, then no conclusion about flying will be reached as *¬penguin(X)* is inconsistent with the available knowledge.

The default reasoning knowledge base given below preserves the default rule and avoids contradictions. The knowledge that X is a bird only implies that it flies if *fly(X)* is not an inconsistent statement (i.e. *¬fly(X)* is not deducible from the rest of the knowledge base).

bird(X): fly(X)/fly(X)
bird(X) ∧ penguin(X): /¬fly(X)
bird(tweety)
penguin(tweety)

Even though non-monotonic logic addresses many weaknesses of monotonic logic, it still suffers from the *multiple extension problem*, in which a proposition and its negation can both be inferred from the knowledge base. Several approaches have been proposed to address the multiple extension problem (Reiter and Crismolo [19], Brewka [2]). For example, priorities can be assigned to the rules using various schemes. In case of conflicting rules the one with the highest priority will be chosen. For example in the previous knowledge base, if rule (2) is assigned higher priority than rule (1), then given only the fact *bird(tweety)* the conclusion *fly(tweety)* is reached. Given both facts *bird(tweety)* and *penguin(tweety)* the applicable rule of highest priority will fire to give $\neg$*fly(tweety)*.

Another proposed method whose intent is to handle inconsistencies induced by uncertainty of knowledge, is the *Plausibility theory*, (Rescher, [20]). Plausibility values are associated with each axiom in the knowledge base. A *maximally consistent subset* of a knowledge base K is a subset C of K such that if D is consistent and $C \subseteq D \subseteq K$ then C=D. In plausibility theory, for the purpose of removing inconsistencies, a maximally consistent subset of the knowledge base with highest plausibility is used in order to extend the knowledge base.

For example, consider the inconsistent knowledge base:

bird(X)⇒fly(X), 0.7
bird(X)∧penguin(X)⇒¬fly(X), 0.9
bird(tweety), 0.99
penguin(tweety), 0.99

where the numeric value next to each axiom indicates the rule plausibility.
The maximally consistent subsets are:

{bird(X)⇒fly(X), bird(tweety), penguin(tweety)}
{bird(X) ∧penguin(X)⇒¬fly(X), bird(tweety), penguin(tweety)}
{bird(X)⇒fly(X), bird(X) ∧penguin(X)⇒¬fly(X), bird(tweety)}
{bird(X)⇒fly(X), bird(X) ∧penguin(X)⇒¬fly(X), penguin(tweety)}

Of the four subsets given above, the second is the most plausibly preferable, since it contains the axioms of the highest plausibility. This consistent subset can be used to uniquely extend the knowledge base by inferring that *tweety cannot fly*.

Another approach to handling inconsistency is employed in the *Truth Maintenance Systems*, (TMS). Rules in a TMS are of the form

$$[S_1,S_2,\dots,S_n],[C_1,C_2,\dots,C_m] \Rightarrow Conclusion$$

where the S_i's are evidence supporting the *Conclusion* and C_i's are evidence against the *Conclusion*. A conclusion is totally believed if all its supporting evidence is believed (as the result of being stated as facts or derived from other rules) and all its contradicting evidence is disbelieved.

For a detailed description of a TMS the reader is referred to Doyle, [5].

Finally, systems which combine the plausibility and truth maintenance approaches have been proposed. For example, in Popchev et al, [17], a contradiction tolerant truth maintenance system, (CTMS), is introduced, where the inference rules are of the form:

If [(S_1, S_2, ...,S_n are true) and (P_1 ,P_2 ,...,P_k are plausible)] then the conclusion C is plausible.

In such a theory, a conclusion and its negation can both be derived with different plausibilities and this is not a contradiction. A contradiction only occurs when both C and not(C) are derived from rules with all the necessary and sufficient conditions being true.

2.10 THE CLOSED WORLD ASSUMPTION, HORN CLAUSES

The *closed world assumption,* (CWA), is a way to represent negation in a logic programming system. Under the CWA, if an atomic formula cannot be derived from the facts and rules in the knowledge base, then it is assumed to be false.

A logic system employing the closed world assumption becomes non-monotonic. Given a knowledge base D with an atomic formula *h* not in the completion of D, (therefore *under CWA, h* is false), it is possible that, when the knowledge base is augmented with new facts or rules (not simply deducible from the knowledge base), then *h* can be proven to be true.

An undesirable effect of the CWA is that it can render knowledge bases inconsistent. Consider the knowledge base D consisting of just the fact $A \vee B$. Since A cannot be derived from the knowledge base, then A does not belong to the completion of D. Similarly, B does not belong to the completion of D. If the CWA is in effect, we can conclude that $\neg A$ and $\neg B$ are true and can be augmented to the knowledge base. But now the knowledge base is inconsistent. As it actually turns out, whenever a clause has two or more positive atomic formulas and at least two of them cannot be deduced from the knowledge base, then the CWA hypothesis renders the knowledge base inconsistent. If we limit ourselves to knowledge bases with clauses of at most one positive literal, then the CWA does not introduce the inconsistency problem.

A clause of at most one positive literal is called a *Horn* clause. The logic programming language Prolog employs the CWA and only Horn clauses, as consistency is of paramount importance. A horn clause with one positive literal can be rewritten in the form of a rule with only one literal at its head. For example, the Horn clause $\neg p(X) \vee \neg q(X,Y) \vee s(X)$ is equivalent to the rule $p(X) \wedge q(X) \Rightarrow s(X)$. It should be noted however that the Horn clause based logic is more restrictive than first order predicate logic. To be able to assure consistency, we sacrifice the full expressive power of first order predicate logic. For example, it is not possible to express the following set of clauses in Horn clause format or Prolog syntax:

$a \vee b \vee c$
$a \Rightarrow b$
$b \Rightarrow c$
$c \Rightarrow a$
and prove the truth of the query $a \wedge b \wedge c$.

2.11 FORWARD, BACKWARD CHAINING

Directed resolution is a form of resolution which can be applied to a knowledge base of Horn clauses. The positive literal of each clause is written either at the start or at the end of the clause and resolution is restricted to the first two literals of the resolvents. Applying directed resolution to Horn clauses with the positive literal written as the first literal gives rise to *backward chaining*. Applying directed resolution to Horn clauses with the positive literal at the end of clauses gives rise to *forward chaining*.

Consider the knowledge base consisting of the fact:

siamese_cat(tom)

and the rules:

cat(X)⇒animal(X) and *siamese_cat(X)⇒cat(X)*.

Assume that the query *?animal(tom)* is issued. The rules can be written in clause form as:

{*¬cat(X)∨animal(X)*, *¬siamese_cat(X)∨cat(X)* } or as
{*animal(X)∨* ¬cat(X), *cat(X)∨¬siamese_cat(X)* }.

Forward chaining is in essence the modus ponens strategy. The knowledge base with the negated query is

$\{\neg cat(X) \vee animal(X), \neg siamese\_cat(X) \vee cat(X), siamese\_cat(tom), \neg animal(tom)\}$.

Directed resolution proceeds by resolving the second and third clauses to get: *cat(tom)*, which resolved with clause one gives *animal(tom)*. Combined with the negated query this gives the empty clause. This amounts to using the fact *siamese_cat(tom)* and rule (2) to imply *cat(tom)*, which combined with rule (1) gives *animal(tom)*. This is forward chaining.

Backward chaining is basically a resolution proof by refutation using the set of support strategy, and corresponds to modus tollens. Every pair of resolved clauses includes the negated query or a clause derived from the negated query. In the example given above, the knowledge base can be written in clause form as:

$\{animal(X) \vee \neg cat(X), cat(X) \vee \neg siamese\_cat(X), siamese\_cat(tom), \neg animal(tom)\}$.

Using directed resolution, the first clause will be resolved with the negated query to give $\neg cat(tom)$, which resolved with the second clause gives $\neg siamese\_cat(tom)$. Finally, $\neg siamese\_cat(tom)$ resolved with the third clause gives the empty clause.

Backward chaining is activated by a query, while forward chaining keeps producing derived clauses until a contradiction is reached proving the theorem-query or until no more resolvable clauses exist, in which case the query is deduced to be false under the CWA.

Logic programming systems can employ either backward or forward reasoning by itself or a combination of the two. Which technique results in the most efficient search of the proof space depends on the structure of the knowledge base. For different modules of the same knowledge base, one or the other technique may be more efficient. For example, consider the following knowledge base:

$horse(X) \Rightarrow mammal(X)$
$cat(X) \Rightarrow mammal(X)$
$eagle(X) \Rightarrow bird(X)$

bird(X) $\Rightarrow$ *animal(X)*
mammal(X) $\Rightarrow$ *animal(X)*
cat(tom)

and the query:

?animal(tom).

In this case, forward chaining is more efficient. Using backward chaining, the subgoal *bird(tom)* will be created (by resolving $\neg$*animal(tom)* with *animal(tom)*$\vee\neg$*bird(tom)*). The subgoal *bird(tom)* will be true if *eagle(tom)* is true. But this is not verifiable. So, the subgoal *mammal(tom)* will be generated (by resolving $\neg$*animal(tom)* with *mammal(X)* $\Rightarrow$*animal(X)*). Using the first rule *mammal(tom)* will be true, if *horse(tom)* is true. But this fails. Using the second rule, the subgoal *mammal(tom)* will be true, if *cat(tom)* is true. Since this is a fact, the backward chaining proof is completed.

If forward chaining is applied to the same knowledge base, the answer to the query is derived much more faster. The fact *cat(tom)* activates the second rule to deduce *mammal(tom)*, which in turn fires the fifth rule to deduce *animal(tom)*.

Consider now the following knowledge base:

cat(X) $\Rightarrow$ *domestic_animal(X)*
domestic_animal(X) $\Rightarrow$ *friendly(X)*
cat(X) $\Rightarrow$ *feline(X)*
feline(X) $\Rightarrow$ *sharp_teeth(X)*
feline(X) $\Rightarrow$ *animal(X)*
cat(tom)

and the query:

?animal(tom).

In this case backward chaining is more efficient. To prove *animal(tom)* we need to prove *feline(tom)* (using the fifth rule), which in turn (using the third rule) gives rise to the subgoal *cat(tom)*, which is true. In contrast, forward chaining wanders down the paths of domestic animals and felines with sharp teeth before reaching *animal(tom)* through the chain cat-feline-animal.

In the first example, the knowledge was structured in a top-down, general to specific fashion and *animal* was the conclusion of a large number of alternate rules, which made backward chaining from *animal* not as efficient as forward. In the second knowledge base, the knowledge structure was from specific to general making backward chaining more efficient. Of course not all knowledge bases are simple hierarchical structures as the ones given above and the problem of deciding whether to use forward or backward chaining is NP-complete for knowledge bases of arbitrary complexity.

SUMMARY

In this chapter, the theoretical foundations of expert systems were introduced. Predicate logic, proof by refutation and resolution were discussed. Various proof search strategies and the theoretical basis of forward and backward chaining were introduced. Various weaknesses of first order logic deductive systems were presented and some efforts to remedy them including nonmonotonic logics were discussed. The declarative logic programming language Prolog is based on the principles of predicate logic, Horn clauses, the closed world assumption and resolution and is introduced in Appendix A.

EXERCISES

1. Express the following sentences in predicate logic format:
   a) Women love roses
   b) Horses and sheep are mammals
   c) No fish except whales and dolphins can breathe air.

2. Translate the following predicate calculus formulas into English statements. Constants have their obvious meanings.
   a) $\forall X\ [apple(X) \Rightarrow (red(X) \vee green(X))$
   b) $\forall X \forall Y \forall Z\ [father(X,Y) \wedge ancestor(Y,Z) \Rightarrow ancestor(X,Z)]$
   c) $\forall X\ \exists Y\ father(Y,X)$

3. Convert the following sentences into clausal form:
   a) $\forall X\ [\exists Y\ (q(X,Y) \wedge r(X,Y))]$
   b) $\forall X\ \forall Y[a(X,Y) \Rightarrow \neg b(X) \wedge \exists Z\ c(Y,Z)]$

4. Determine if following pairs of expressions unify with each other

and if so give the most general unifier.

a) loves(X,brother(X)) and loves(Y,Z).

b) married(joe, wife(X)) and married(Y, husband(Y)).

5. If a student is a ACM member then he receives a discount in the membership fee. Mike is an ACM member, but receives no discount.

   (i) Express this knowledge in predicate logic format.
   (ii) Convert the logic formulas into clause form.
   (iii) Using resolution prove that Mike is not a student.

6. Solve the missionaries and cannibals problem using resolution.

   This problem is often used in AI to demonstrate various techniques and is stated as follows. A group of n missionaries and n cannibals is on the East side of a river and there is one boat available to them. The boat can hold at most two people and must have at least one person to do the rowing. If at any time, on either side of the river the cannibals outnumber the missionaries, then the cannibals will overpower and eat the missionaries. Express the problem in predicate logic format and using resolution show that there is a way to get everybody on the West bank of the river without having anyone eaten.

REFERENCES

[1] Bratko, I., *Prolog Programming for Artificial Intelligence*, Addison-Wesley, 1986.

[2] Brewka, G., *Nonmonotonic Reasoning: From Theoretical Foundation towards efficient computation*, Ph.D thesis, University of Hamburg, 1989.

[3] Chang, C.L., Lee, R.C.T., *Symbolic Logic and Mathematical Theorem proving*, New York, Academic Press, 1973.

[4] Clocksin, W., Mellish, C., *Programming in Prolog*, Springer Verlag, 1982.

[5] Doyle, J., A truth maintenance system, *Artificial Intelligence*, 12, pp.231-272, 1979.

[6] Frege, G., *Begriffschrift eine der Arithmetischen Nachgebildete Formelsprache des reinen Denkens*, Halle:Louis Nebert, 1879.

[7] Green, C.,"Theorem proving by resolution as a basis for question answering", *Machine Intelligence*, 4, 1969.

[8] Kowalski, R.A. and Kuehner, D., "Linear resolution with selection function", *Artificial Intelligence*, 2, 1971.

[9] Kowalski, R.A., "Predicate logic as a programming language", *Proceedings of the IFIP Congress*, 1974.

[10] Loveland, B.W., "Quarter century review", in Bledsoe, W.W. and Loveland, D.W., (eds), *Automated theorem proving: After 25 years*, Providence, R.I.,American Mathematical Society, 1984.

[11] McCarty, J., "Circumscriptionm: A form of non-monotonic reasoning", *Artificial Intelligence*, 13, 1-2, 27-39, 1980.

[12] McDermott, D., Doyle, J., "Non-monotonic Logic", *Artificial Intelligence*, 13, p. 41-72, 1980.

[13] Meltzer, B., "Some notes on resolution strategies" in Michie, D, (ed.), *Machine Intelligence,* 3, Elsevier, N.Y. , 71-75, 1968.

[14] Newell, A., Shaw, J.C., Simon, H.A., "Empirical explorations with the Logic Theory Machine: A case study in heuristics", In: *Computers and Thought*, Feingenbaum, E. A., Feldman, J., (Eds.), McGraw Hill, New York, 1963.

[15] Nikolopoulos, C. "An intelligent search strategy for logic programs", *Proc. of Fourth Intern. Confer. on Symbolic and Logical Computing*, 1989, pp. 307-318.

[16] Nilsson, N., J., *Principles of Artificial Intelligence*, Springer Verlag, 1982.

[17] Popchev, I., Zlatareva, N., Mircheva, M., "A truth maintenance theory: An alternative approach", ECAI '90, *Proceedings of the 9th European Conference on Artificial Intelligence*, August 1990, pp. 509-514.

[18] Reiter, R., "On closed world data bases", *Logic and Databases*, Plenum,Eds. Galaire, H. and Minker, J., 1978.

[19] Reiter, R., "Non-monotonic Reasoning", *Annual Review of Computer Science*, 2, 1987, pp. 147-186.

[20] Rescher, N., *Plausible Reasoning: An Introduction to Theory and Practice of Plausibility Inference*, Van Gorcum, 1976.

[21] Robinson, J.A., "A machine oriented logic based on the resolution principle', *Journal of the ACM,* 1965.

[22] Smith, D.G., Genesereth, M.R., Ginsberg, M.L., "Controlling recursive inference", *Artificial Intelligence*, 30(3), 343-389, 1986.

[23] Wos, L. et al., *Automated Reasoning:Introduction and Applications*, Englewood Cliffs, NJ, Prentice Hall, 1984.

CHAPTER 3

KNOWLEDGE REPRESENTATION

3.1 REPRESENTATION OF KNOWLEDGE

Knowledge acquisition encompasses the processes of knowledge elicitation and knowledge representation. *Knowledge elicitation* deals with the issues involved in acquiring the necessary knowledge and constructing a conceptual model of this knowledge in order to solve the problem at hand. *Knowledge representation* is the process of choosing an appropriate knowledge representation formalism, in which the acquired knowledge is encoded so that it becomes amenable to computer manipulation. The amount of competence, which an expert system exhibits in solving problems, depends on the quantity and quality of the knowledge obtained as a result of the knowledge elicitation process. The efficiency, with which problems are solved, depends among other things on the knowledge representation formalism chosen to encode this knowledge. This Chapter mainly deals with the topic of knowledge representation. Knowledge elicitation is covered in Chapter 4.

In general, a good knowledge representation formalism must (Winston, [27]):

-make important things explicit
-facilitate the processing of the knowledge within the system
-be concise
-be transparent to human users and facilitate machine/user interaction
-suppress unnecessary detail

Whether a knowledge representation formalism meets the above criteria or not depends to a great extent on the type of problems which we intend to solve. The same knowledge representation formalism can meet the criteria given above and be appropriate, or not meet them, depending on the type of queries that are to be asked of the knowledge base. Some types of representations can represent certain types of knowledge more efficiently than others, even though the representation schemes may be of equivalent *epistemological value* (ability to express the same types of knowledge, a term introduced by McCarthy, [12]). A knowledge representation structure, which is natural to the problem domain, can greatly facilitate the building of the system's development tools, explanation facilities and interfaces. A representation, which does not match the problem domain, can introduce unnecessary obstacles in the implementation of these components.

For example, consider the following two representations of natural numbers. A natural number can be represented:

i) in its decimal form

or

ii) as a vector of its prime factors and their exponents, ($prime_1$, $exponent_1$,..., $prime_n$, $exponent_n$). For example, 360=(2,3,3,2,5,1), since $360 = 2^3 \times 3^2 \times 5$.

Clearly, the two representations are isomorphic and of equal epistemological value, since one can be mapped onto the other, but which representation is better, depends on which problem is to be solved. If the problem is *"given a number, decide if it is a prime"*, then the second representation is better, since in order to answer the query, all that has to be checked in the vector implementation is whether it is a two dimensional vector and the second coordinate is equal to 1. In this case, the first representation does not make important things explicit and does not facilitate computation. But if the problem to be solved is *"given a number, is it greater than 1000?"*, then the first representation is better. The second representation does not facilitate computation, is not concise and contains a lot of unnecessary detail not relevant to the problem.

The reader familiar with Data Structures and algorithms in procedural programming will find that there is an analogy here with choosing the proper data structure to implement an Abstract Data Type. Which data structure is more suitable and efficient depends to a great extent on what types of transactions are to be processed on the data.

After the domain and control knowledge is encoded in a knowledge representation formalism suitable for the problems that are to be solved, it has to be maintained and updated whenever necessary. When new knowledge structures are added, they have to fit the overall knowledge structure and avoid rendering the knowledge base contradictory. In addition to improving efficiency, the choice of a proper knowledge representation scheme can also make updating the knowledge base and the verification process easier.

For example, consider the knowledge that "*every plane has exactly two pilots*", which can be written in predicate logic as:

$$\forall X\ plane\ (X) \Rightarrow ([\exists Y\ \exists Z\ (pilot(Y,X) \wedge pilot(Z,X) \wedge \neg equal(Y,Z)) \wedge (\forall W\ pilot(W,X) \Rightarrow (equal(W,X) \vee equal(W,Z)))]$$

The same knowledge can be expressed much more easier in set theoretic notation as:

$$cardinality(pilots(plane(X))=2.$$

The set theoretic representation is more concise. An obvious advantage of a concise representation is that it is easier to modify. For example, to state that "every plane has at most two pilots", all we have to do in the set theoretic notation is change the "=" to a "<=", while it is not immediately obvious how to update the predicate logic formula to accommodate the modification.

Answers to posed queries are derived by manipulating or reasoning using the knowledge base. Different knowledge representation techniques support different types of inference processes. The reasoning process does not necessarily have to be based on logical inference. For example, statistical models and connectionist models of inference have been developed. Which type of inference method can be employed depends on the choice of the knowledge representation. In general, even if two representations are of equivalent expressive power, (like rules, frames and logic), they may differ on how efficiently they will handle certain types of inference. For example, rules may be better at entailment, while frames are better at taxonomic classification and default inheritance. In addition, when different types of knowledge, (such as taxonomic, causal or temporal), are present in the same domain, a combination of representation schemes in a hybrid scheme may be needed to express the various knowledge types more efficiently.

Finally, in order to answer queries about the problem domain efficiently, techniques have been developed to retrieve the part of the knowledge base relevant to the query. Since knowledge bases representing real world problems can be extensively large, the brute force approach, of trying all possible alternatives until a solution is reached, leads to combinatorial explosion. Retrieval methods depend on the representation model the choice of which can thus have an additional impact on the efficiency with which queries are answered.

The main knowledge representation formalisms proposed in the literature include rules, semantic nets, frames, scripts, petri nets, and the object oriented paradigm.

3.2 RULES

First generation expert systems were predominantly based on production rules. For example, MYCIN used rules to represent its knowledge base, and R1/XCON mainly contained rules, even though to a lesser extent it also used specialized representations to solve board configuration problems.

Rules (or production rules) represent knowledge in the form of:

IF(condition) THEN (conclusion or action).

The specific syntax varies. For example, the knowledge that "if a car does not start and the head lights do not go on, then the battery is dead", can be expressed in Prolog as:

diagnosis(battery_dead):-car_start(no), lights_work(no).

The same rule in the expert system shell Exsys reads:

Rule 1
IF car_start is no and lights_work is no THEN diagnosis is battery_dead;

while in OPS5 (a forward chaining production systems language) the rule looks like:

```
(p rule1
        (car ^start no)
        (lights ^work no)
-->
(make Diagnosis ^Type battery-dead))
```

A more detailed description of the expert system shell EXSYS is contained in Appendix B. A description of the principles behind production systems is given in the next section.

As explained in Chapter 2, there are two ways to perform reasoning in a rule based system. In the data driven reasoning model, (forward chaining), inference proceeds from the condition to the conclusion and is initiated by the facts or data. In the goal driven model, (backward chaining), the search for the verification of a specific goal is conducted from the conclusion to the hypothesis (to prove conclusion, it is sufficient to prove hypothesis, etc.). The expert systems Dentral and R1/XCON are examples of forward chaining systems, while MYCIN basically used backward chaining. For the theoretical basis and comparison of forward and backward chaining, see Section 2.11.

3.2.1 Production systems and the Rete matching algorithm

3.2.1.1 Production systems

A typical production system contains a *production memory*, a *working memory* and an *inference engine*. The production memory (knowledge base) consists of production rules and the working memory holds the facts (original or deduced from the knowledge base). A rule is a candidate for firing when there is a *consistent match* of its conditions to elements in working memory. A match of the conditions of a rule to facts is *consistent* if each variable, which is shared by more than one condition, is bound to the same value. Inference in a production system is conducted in *match, select and act* cycles. In the match cycle, the set of rules are checked against the facts in working memory and the rules all of whose conditions are consistently matched constitute the *conflict set*. Rules in the conflict set are stored in the *agenda*. The same rule may appear in the agenda multiple times, as its conditions may be consistently matched by different sets of working memory elements. In the selection phase, a selection or *conflict resolution strategy* is employed to select which rule among those contained in the conflict set should be fired. In the act phase of the cycle, the selected rule is fired. The firing of a rule may remove, add or modify elements in the working memory. Changes in the working memory produce new consistent matches of the conditions in the rules, which in turn results into changes in the agenda. Looping through the *match-select-act* cycles repeats itself until the agenda becomes empty or some other stopping criterion is met (for example, when the number of rules fired reaches a maximum).

Production rules do not communicate with each other, in contrast to parameter passing in procedural environments. Instead, the rules only interact with the contents of the working memory, waiting eligibility to fire and when a rule fires, it posts its action on the working memory bulletin board, making it accessible to all the other rules.

General production system languages include OPS5 and CLIPS. Object oriented production systems environments have also become available, like COOL (object oriented extension of CLIPS) and JESS. JESS (Java Expert System Shell) is written in SUN's Java language and it further extends the features of CLIPS by providing the ability to use scripting applets and interface with the Web. OPS5 is a general production systems language, which has been used to implement a number of large expert systems. It was originally developed in conjunction with the R1/XCON expert system used by DEC to configure computers. In OPS5, a *time tag* is associated with each working memory element. The time tag indicates when the element was last modified or added into the working memory. The larger the time tag, the more recent the element. Time tags are used by the conflict resolution strategy. The programmer has a choice of two conflict resolution strategies, LEX and MEA, which are based on the recency of matching facts (time tags) and generality of rule conditions. LEX is simpler than MEA, while MEA provides more execution control than LEX. CLIPS, which is written in C, was developed by NASA and is similar to OPS5.

3.2.1.2 The Rete matching algorithm

It would clearly be very inefficient to recompute all possible consistent rule matchings each time through a match cycle. When a rule fires, most of the working memory elements and therefore their effect on the rules remains unchanged. RETE, (Forgy, [7]), is an efficient match algorithm which was developed in conjunction with OPS5. It attempts to speed up execution of production systems by reducing the time spent in the match cycle. RETE is being used by production systems like OPS5, ART and CLIPS, and by the most recent expert system shell JESS. The interested reader is referred to appendix C for references to other systems based on the Rete algorithm and recent extensions of Rete such as Rete++.

It has been estimated that up to 90% of a production system's run time is spent on performing repetitive pattern matching between the rule set and the working memory elements. In RETE, the information about the matches of conditions to working memory elements is stored in a network structure. The nodes of the network correspond to individual condition elements. Each node has two sets associated with it. The first of these sets contains all the working memory elements that the condition node matches. The second set contains combinations of working memory elements and the bindings which produce a consistent match of the conditions that chain up to the node condition. This configuration allows the matcher to avoid repetitive testing of all the rule conditions in each cycle, and only check those nodes affected by a newly inserted or modified fact.

To give an example of how this works, consider the following two rules:

If a(X,1) and b(X,Z) THEN g1(X,Z)

and

If a(X,2) and b(X,Z) THEN g2(X,Z)

The corresponding network structure is shown in Figure 3.1. There is a starting node and a node for each of the rule conditions and conjunctions of conditions. The conjunctive nodes of outdegree 0 correspond to production rules. An element, which is to be added or deleted from working memory, is represented by a token. A token is initially deposited at the start node and traverses the network. The network arcs are labeled with bindings or relations among the variables of the condition predicates of the node at the arc origin. A token (working memory element) propagates through an arc only if its arguments satisfy the relation at the arc label. In a conjunctive node, a comparison of the tokens deposited there is performed. If there are tokens matching each of the conditions in the conjunction and the match is consistent among the conjuction variables, then that rule is added to the conflict set.

In the example given in Figure 3.1, suppose that the working memory is empty (therefore all condition nodes are also empty). Assume that the fact a(3,1) is added into the working memory. It will then be deposited in the node labeled a(X,Y) and will propagate through the arc labeled with the constraint Y=1, but not through the arc labeled Y=2. Node labeled a(X,1),b(X,Z) holds a matching token, a(X,1), but no token matching b(3,Z), so the first rule is not added to working memory. The nodes corresponding to b(Y,Z) and a(X,2),b(X,Z) are still empty. Assume

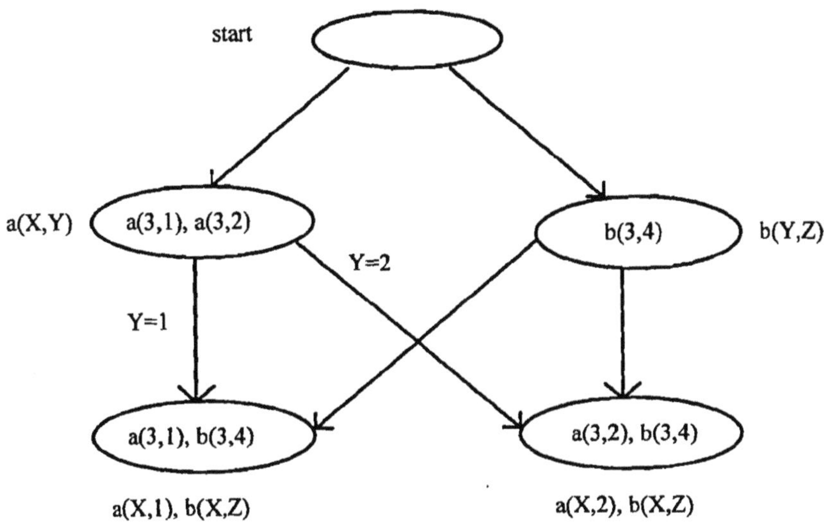

Figure 3.1 A network structure for the RETE matching algorithm.

that the fact b(3,4) is added next into working memory. It will go through node b(Y,Z) and it will also be deposited in the nodes labeled a(X,1),b(X,Z) and a(X,2),b(X,Z). The first coordinate of b(3,4) matches the first coordinate of token a(3,1) which is already in node a(X,1),b(X,Z), so rule 1 is added in the conflict set. Assume next that the fact a(3,2) is added into the working memory. Then token a(3,2) will be deposited in the node labeled a(X,2),b(X,Z), which together with b(3,4) already in that node will have the effect of placing the second rule in the conflict set. When elements are deleted from the working memory, corresponding tokens are similarly propagated through the net, deleting their copies from nodes and deleting corresponding rules from the conflict set instead of adding them. Clearly, in the procedure given above, no repetitive search through the rule set or the element set takes place. This results in a very efficient match algorithm.

3.3 SEMANTIC NETWORKS

3.3.1 Definitions, examples

A semantic network, (Quillian, [16]), is a labeled directed graph representing concepts, objects, their properties and their relationships. In

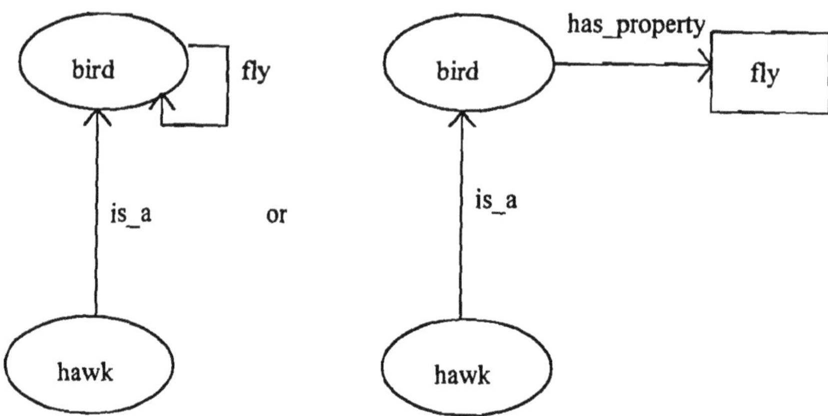

Figure 3.2 A semantic net for: "A hawk is a bird and birds can fly".

the graphical representation of a semantic network, the nodes are oval shaped and they represent concepts, objects, or situations. The arc labels are intended to describe the relationship between two concept nodes. A self arc to a node describes a property of that node. Alternatively a property node can be used in the shape of a rectangle to hold a property name and a "has_property" arc can be used to connect an object to its property. The inference algorithms associated with a semantic network are basically graph traversal algorithms. In Figure 3.2, the knowledge that "a hawk is a bird and birds can fly" is expressed as a semantic network in two equivalent ways. For more formal approaches to the definition of the semantics and the syntax of semantic networks, the reader is referred to Brachman, ([2]).

While it is easy to see how to represent binary relations using semantic networks, it is not so apparent how nonbinary relations should be represented. For example, consider the knowledge "Mike is Christian's and Basil's academic advisor. He advised Christian to take a statistics course and Basil a calculus course". The relationship "advisor" is ternary, since it involves the advisor, the advisee and the course recommended by the advisor to the advisee. A methodology for representing nonbinary relations was introduced in ([21]) which works as follows. If a node owns a set of arc relations, which are related in one unit, (advisor owns advisee and course), a node called the *"situation"* or *"case"* node is introduced between the owner node and the arcs. The particular set of property arcs referring to the same object, are arcs with origin at an instance of this

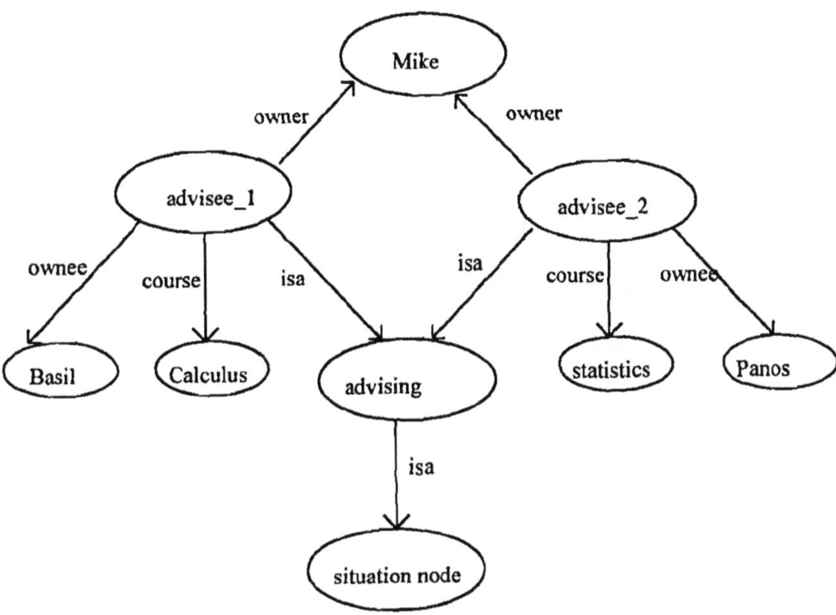

Figure 3.3 The semantic network for the nonbinary relationship "advising".

situation node. For the reader familiar with hierarchical databases, it should be recognizable that the proposed solution is similar to how hierarchical databases treat many to many relationships among data. Using this technique, the knowledge given above can be expressed as a semantic network by introducing a situation node for the advisor relation. The representation is shown in Figure 3.3.

3.3.2 Semantic nets in Prolog (optional)*

A semantic network knowledge structure can be converted quite easily into a Prolog rulebase. For example, the net of Figure 3.2 can be represented by the following Prolog rules, where the last three rules implement inheritance of properties:

isa(hawk,bird).
has_property(bird,fly).
has_property(X,Z):-chain_isa(X,Y),has_property(Y,Z).

```
chain_isa(X,Y):-isa(X,Y),!.
chain_isa(X,Y):-isa(X,W),chain_isa(W,Y).
```

3.3.3 Advantages of semantic networks

One of the main advantages of using semantic networks to represent knowledge is their intuitive, easy to comprehend nature. Their ability to easily represent relationships between concepts and especially hierarchies, makes them particularly suitable for representing taxonomic knowledge and solving classification problems in which reasoning is based on taxonomies. In addition, the semantic net formalism permits *property inheritance*. For example, in Figure 3.2, from the relations that a hawk is a bird and birds can fly, we can conclude that a hawk can fly. The natural way, in which inheritance is implemented in semantic networks, makes their reasoning process very efficient for *inheritance* and *recognition* type of queries or combinations of such queries (an *inheritance query* involves finding a property value for a given concept and a *recognition query* involves finding a concept which possesses certain properties).

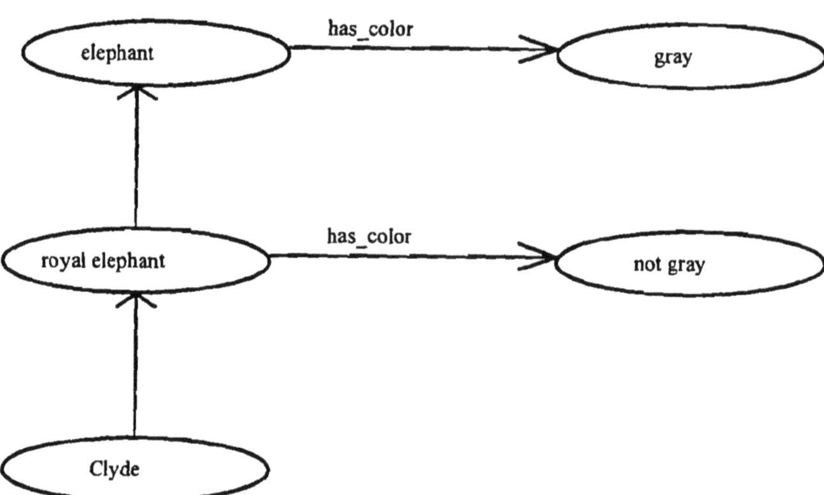

Figure 3.4 The exception problem in semantic networks.

3.3.4 Exception and multiple inheritance

Even though inheritance is naturally implemented in semantic nets facilitating the query types described above, it can also lead to the *exception* and *multiple inheritance* problems.

Consider for example the semantic network of Figure 3.4, which is intended to express the knowledge "Most elephants are gray. Royal elephants are elephants. Most royal elephants are not gray. Clyde is a royal elephant" (Brachman, [3]). The inheritance mechanism can derive that the color of Clyde is both gray and not gray. But, royal elephants should not inherit the color property of elephants; they are an *exception.* One way to resolve conflicts created due to the "exception problem" is by inheriting the property value of the closest ancestor of the queried concept. As a consequence, Clyde is not gray.

The *multiple inheritance problem* is depicted in Figure 3.5. This figure expresses the knowledge: "Mike is a student and a soldier. Most students have long hair. Most soldiers have short hair". Mike can inherit two mutually exclusive properties, having long and short hair at the same time.

Several approaches have been proposed to solve the multiple

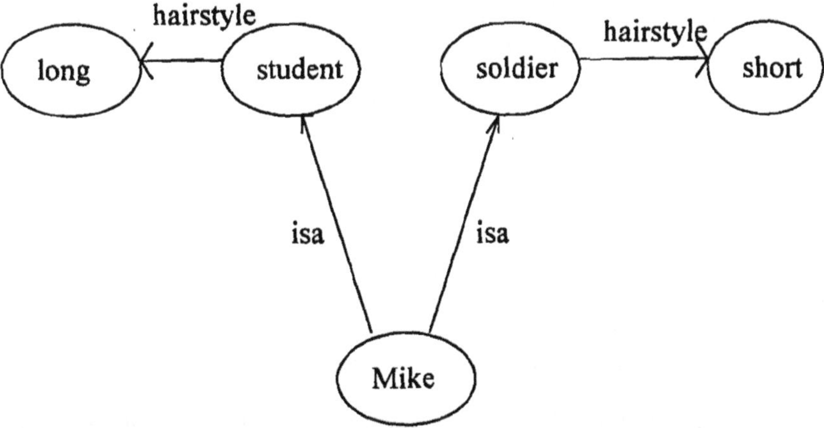

Figure 3.5 Multiple inheritance problem in semantic nets. What is the hairstyle of Mike?

inheritance and exceptions problems in semantic nets, such as default logic, (Etherington, Reiter, [5]), inferential distance ordering, (Touresky, [26]), and evidential reasoning, (Shastri, [20]). In the next section, we briefly examine the last approach.

3.3.4.1 Evidential reasoning

The *evidential reasoning* approach seems to be a very effective method for solving multiple inheritance and exceptions problems (Lim and Cherkassky, [11]). In evidential reasoning, evidential information in the form of relative frequencies is employed in order to choose the most likely answers. For example, assume that there is a total of 100 elephants, out of which 40 are royal elephants, 70 are gray and 30 are not gray. Then the evidence which supports the proposition that "a royal elephant is gray" can be approximated by :

(number of royal elephants×number of gray elephants) / (number of elephants) =40×70/100= 28

The evidence in a royal elephant not being gray is:

(number of royal elephants × number of not gray elephants)/(number of elephants) = 40×30/100 =12

So, if we know that Clyde is a royal elephant, based on the relative frequency evidence, we would conclude that Clyde is gray.

3.3.5 Conceptual graphs

A *conceptual graph* is a directed, connected graph whose nodes and arcs represent *concepts*, *concept instances* (called *referents*), *conceptual relations* and *actors*. The first three represent declarative type of knowledge, while an actor represents procedural knowledge. An *actor* is a process whose input and output are concepts and which can change the referents of the output concepts based on the input concepts.

A conceptual graph can be represented by either a graphical or a linear notation. In linear notation, the conceptual graph is a string of characters with concepts and referents being enclosed by square brackets [ ], relations by parentheses ( ), and actors by single brackets <>. There are many types of referents, including:

1. The *generic* type, denoted by [C:*x] with interpretation: there is an x which is an instance of concept C.
2. The *universal* type, denoted by [C:∀x] with interpretation: for every x, x is a member of C
3. The *proper name* type, denoted by [C:word] with interpretation: word is the name of a member of C.
4. The *negative* type, denoted by [C:¬] with interpretation: for all x, x is not a member of C.

Conceptual graphs can be used to define the syntax and semantics of semantic networks. For example, consider the conceptual graph describing the advisor-advisee relationship of the previous section and let *evaluate* be an actor which calculates the total number of credit hours that a student takes by summing up the credit hours per class. In linear form, this knowledge can be represented in a conceptual graph as follows:

```
[advising]-
        ←(instance)[advisee1]-
                →(owner)[advisor:mike]
                →(ownee)[advisee:basil]
                →(hours)[total credit hours: *x]
                        ←<evaluate>-
                                ←(course)[calculus: *y1]
                                ←(course)[philosophy: *y2]
                                x=y1+y2;

        ←(instance)[advisee2]-
                →(owner)[advisor:mike]
                →(ownee)[advisee:christian]
                →(hours)[total credit hours: *y]
                        ←<evaluate>-
                                ←(course)[physics: *y1]
                                ←(course)[statistics: *y2]
                                y=y1+y2;
```

Conceptual graphs can also be used to define a set of *primitive blocks*, which can form the basic components for building more complex semantic net structures. For a composition theory of how complex semantic nets can be composed from a set of primitive "conceptual graphs", the reader is referred to Sowa, ([24]).

In addition, the basic notion of a conceptual graph has been extended to incorporate a type of node called *demon* (giving rise to *temporal* conceptual graphs) and *constraint* (or *actor*) *overlays*, (Pfeiffer, [15]). A demon is represented by double brackets and has the ability to assert or retract concepts. When all the input concepts to a demon node have been asserted, then the demon is activated, asserts each of its output concepts with referents and retracts its input concepts. Constraint overlays attach constraints to concept nodes and describe changes to the model state.

Generalized conceptual graphs with demons and actor overlays have been used as a formalism for representing the conceptual model of a problem domain which is produced by the task based specification methodology of second generation expert systems (see Chapter 4 on knowledge acquisition).

3.4 FRAMES

3.4.1 Basic concepts

A *frame* is a collection of *"slots"* which are logically related by their correspondence to the attributes of the same concept or object. A slot can hold an individual value, (analogous to fields in a record), or a set of values. In addition, and unlike records or traditional relational database table attributes, a slot can also hold other objects, a set of objects, algorithms for performing various tasks (called *demons*), scripts, pointers to other frames, as well as other complex object structures generated by combining the above elements (Minsky, [13]).

For example, the concept of an "automobile" can be represented as a frame with the following associated slots:

model_type, engine_type, sunroof, license_plate_number, mileage, manufacturer, refuel, buy_a_car.

The range of the slot *model_type* is {sedan, minivan, truck, bus, wagon}, the range of *engine_type* is {4_cylinders, 6_cylinders, 8_cylinders}, the range of *sunroof* is {yes, no}, the range of *manufacturer* is {ford, chevrolet, isuzu}. An *if-needed* demon attached to a slot is an algorithm which is called upon when there is a need to determine the value to be stored in an empty slot. An *if-added* demon attached to a slot is activated

```
Frame_name: automobile
Model: range=[sedan, minivan, truck, bus, wagon]
        default_value=sedan
Engine: range=[4cylinder, 6cylinder, 8cylinder]
        default_value=4cylinder
Sunroof: range=[yes, no]
        default_value=no
Manufacturer: range=[ford, chevrolet, mercedes]
Mileage: If-needed(miles_traveled/gallons_used)
Buy_a_car: [go_dealer, choose_car, bargain, get_loan, sign_papers]
```

Figure 3.6 A frame structure.

whenever a value is added into that slot. In our example, the slot *"refuel"* has a boolean range and holds an *if-needed* method which calculates the need to refuel or not based on the fuel level. For example, if the fuel level is less than 1/4 full, then refuel. After the if-needed method returns a value for the slot "refuel", an if-added method can be activated which resets the value of the "refuel" slot to blank. The slot *buy_a_car* contains the *script* (go_dealer, choose_car, bargain_price, get_loan, sign_papers, drive_off). A *script* is a sequence of events or description of certain situations, which are expected to be encountered in a certain environment. A graphical representation of the automobile frame is shown in Figure 3.6.

The fact that frames can be organized in frame hierarchies from more general to more specific concepts, makes them suitable for those problems where there is a built-in natural taxonomy of concepts. Figure 3.7 shows a frame hierarchy. The ISA link indicates that the frame *sedan* is a kind of automobile. The *sedan* frame is a *subframe* of the frame automobile. An ISA link implies an inference. In this example, the implied inference is that "if an object X is a sedan then X is an automobile". Observe that in the car_1 frame, the slot (# of miles in last trip, # of gallons used) takes a complex structure as a value. Namely, a pair of numbers corresponding to the miles driven and gallons used, as well as a method called "update_mileage". This is an if_added method in the sense that if the number of miles and gallons are added in the slot then the method will activate and calculate and add the value of mileage

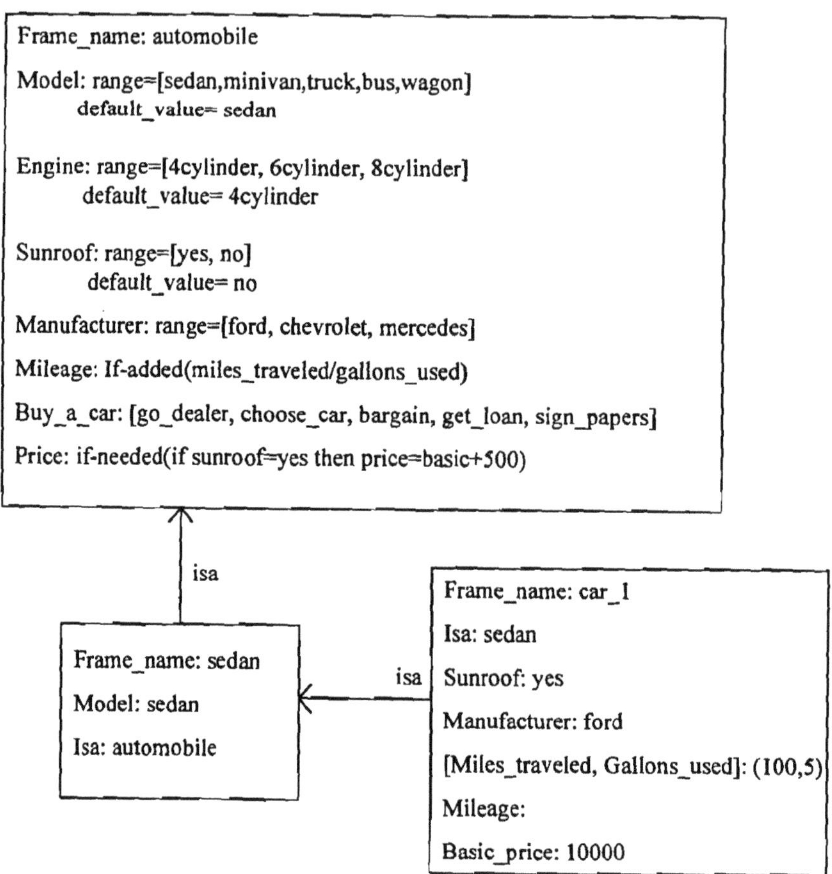
Frame_name: automobile
Model: range=[sedan,minivan,truck,bus,wagon]
default_value= sedan
Engine: range=[4cylinder, 6cylinder, 8cylinder]
default_value= 4cylinder
Sunroof: range=[yes, no]
default_value= no
Manufacturer: range=[ford, chevrolet, mercedes]
Mileage: If-added(miles_traveled/gallons_used)
Buy_a_car: [go_dealer, choose_car, bargain, get_loan, sign_papers]
Price: if-needed(if sunroof=yes then price=basic+500)
isa
Frame_name: sedan
Model: sedan
Isa: automobile
isa
Frame_name: car_1
Isa: sedan
Sunroof: yes
Manufacturer: ford
[Miles_traveled, Gallons_used]: (100,5)
Mileage:
Basic_price: 10000

Figure 3.7 A frame hierarchy.

in the mileage slot.

A frame *inherits* slot values from other frames higher up in the hierarchy. Inheritance is an essential feature of frame based knowledge representation systems. For example, in the frame structure of Figure 3.7, the answer to the query "What is the engine type of a sedan?" should be "4-cylinders". Since the sedan frame has no value for the slot "engine type", its ancestors in the hierarchy will be searched and the *sedan* frame will inherit the value of engine type from the first ancestor frame which owns such a value. Default inheritance can be overridden by filling in the value of a slot.

Various specialized frame languages have been proposed to facilitate the creation and manipulation of frame structured knowledge, (for example, KRL by Bobrow, [1], KLONE by Brachman, [2]). They provide operators to create, update and extract information from frames.

3.4.2 Frames in Prolog (optional)*

We show below how to represent the frame hierarchy of Figure 3.7 in Prolog and write some of the Prolog predicates needed to manipulate the frame structure. Each frame can be represented in Prolog syntax as a collection of facts of the form: *frame_name(Slot,Slot_value)*. The predicate *add(Frame, Slot, Value)* is used to add a new slot and its value in frame *Frame*. The predicate *ask(Frame, Attribute, Value)* retrieves the *Value* for slot *Attribute* in frame *Frame*, using the frame hierarchy, if necessary.

```
automobile(model, sedan).
automobile(engine_type, 4_cylinder).
automobile(sunroof, no).
automobile(mileage, unknown).
sedan_type(model, sedan).
truck_type(model, truck).
isa(sedan_type, automobile).
isa(truck_type, automobile).
isa(car_1, sedan_type).
car_1(sunroof, yes).
car_1(manufacturer, ford).
car_1(license_plate, us7554).
car_1(basic_price, 10000).
```

```
(1) ask(Frame, Attribute, Value):- M=..[Frame, Attribute, Value],
                    call(M), !.
(2) ask(Frame,Attribute,Value):-
                    isa_chain(Frame,G),M=..[G,Attribute,Value],call(M),!.
(3) ask(Frame, Attribute, Value):-
                    Attribute=price,find_price(Frame,Attribute,Value).
(4) isa_chain(X,Y):-isa(X,Y).
(5) isa_chain(X,Y):-isa(X,Z), isa_chain(Z,Y).
(6) add(Frame, Slot, Value):- not(var(Value)),
                         L=..[Frame, Slot, Value], asserta(L),
                         (Slot=miles_gallons, update_mileage(Frame));true.
```

```
% if-added method
(7) update_mileage(Frame):-K=..[Frame,miles_gallons, [M, G]],
                call(K), ML is M/G, L=..[Frame,mileage,W],
                retract(L), N=..[Frame,mileage,ML],asserta(N)).
% if-needed method
(8) find_price(Frame,price, Value):-
                ask(Frame,basic_price,V),ask(Frame,sunroof, Answer),
                Answer=yes, Value is V+500,!.
(9) find_price(Frame,price,Value):- ask(Frame,basic_price,V),
                Value is V.
```

Suppose that we want to know if car_1 has a sunroof. Then the following query is issued on the knowledge base:

```
?-ask(car_1,sunroof,Answer).
Answer=yes
```

The value of Answer was derived by using rule (1). M is bound to *car_1(sunroof,Answer)*, which is then issued as a sub-query. It unifies with the corresponding fact by binding Answer to yes.

Consider next the query:

```
?-ask(car_1,model,Value).
Value=sedan
```

Using rule (1), M is again bound to car_1(model, Value), which is then issued as a query. But in this case, there is no fact to match it, so rule 1 fails. The information about the car model is stored in a frame higher in the hierarchy. Similarly, rule 2 fails because the value for Attribute is not equal to *price.* When rule (3) is used, the subgoal *isa(car_1,G)* succeeds by binding G to sedan_type. M is bound to *sedan_type(model, Value)*, which is then issued as a query. The query succeeds by matching a fact which binds *Value* to *sedan.*

Consider next the query:

```
?-ask(car_1,engine_type,Value).
Value=4cylinder
```

Rules (1) and (2) fail, so rule (3) is activated. The subgoal *isa(car_1,G)* succeeds by binding G to sedan_type. The query *sedan_type(engine_type, Value)* is issued but fails. So, backtracking takes place to the subgoal isa(car_1, G) and alternate bindings are searched for. Rule (4) is used and the query isa(car_1,G) succeeds with the bindings Z=sedan and G=Y=automobile. Back to rule (3), where the subgoal automobile(engine_type,Value) is issued and matches a fact by binding Value to 4cylinder.

As another example, consider the query:

?-ask(car_1,price,Value)
Value=10500.

Rule (1) fails, so rule (2) is used and the subgoal *find_price(car_1,price,Value)* is pursued. Rule (7) is tried, the subgoal car_1(price,Value) is issued but fails. So rule (8) is attempted. The subgoal *car_1(basic_price,V)* is issued which unifies with a fact to yield V=1000. Next, the subgoal *car_1(sunroof, Answer)* is issued and becomes true for Answer=yes, so *Value* becomes equal to 10500.

As a last example, showing the use of the if-added method, consider the query:

?-add(car_1, miles_gallons, [100,5]).

Using rule (5), the new fact *car_1(miles_gallons,[100,5])* will be created and the method *update_mileage* will be called upon. The method calculates the mileage and assert the new fact *car_1(mileage,20)*.

3.5 OBJECT ORIENTED SYSTEMS

3.5.1 Object oriented concepts, examples

The object oriented paradigm has emerged from developments in the disciplines of programming languages, artificial intelligence and databases. In programming languages, the object oriented concepts were the results of extensions of the concepts of abstract data types, information hiding and encapsulation. In artificial intelligence, Minsky's theory of frames was used for knowledge representation and is analogous to object oriented programming. Finally in databases, semantic data modelling research evolved to object oriented database concepts.

We describe below a set of characteristics, whose presence is necessary in order to label a system as "object oriented". These features are encapsulation, information hiding, methods, messages, inheritance, overloading, polymorphism and late binding.

Objects, Classes:

Each real world entity is modeled by an *"object"*. Each object has existence independent of its value and is assigned a unique *object identifier*, (concept of "object identity"). Each object contains a set of instance *attributes* (slots, instance variables) and *methods* (functions or procedures). Flow of control in object oriented systems takes place by sending *messages* between objects. Data abstraction prevents an object's data from being directly manipulated. To access an object's attributes and values a message must be sent to the object. Message sending can be viewed as an indirect procedure call. Instead of naming a procedure to manipulate an object, one sends the object messages to which the object responds using its "methods".

Class Inheritance:

A *"class"* is a group of objects that share the same structure and behavior. Each object is an instance of some class. A class which is a specialization of one or more classes, is called a *subclass*. A subclass inherits attributes and methods from its *superclasses*. This hierarchical structure, (inheritance), enhances the usability of software and reduces redundancies.

Overloading and polymorphism:

In *overloading*, the same operator is allowed to represent multiple operations on different data types. For example, the operator "+" can be used to add reals or integers. *Polymorphism* extends the principle of overloading by allowing methods to operate on different types. For example, a method "sort", which when used in a message can sort either an integer or character array, will be polymorphic.

Late binding:

The system does not bind an operation name to the corresponding source code at compile time but at run time.

Overriding:

An operation of most general type which is defined in the hierarchy lattice can be redefined in some subclasses, overriding the default inheritance of the general type.

Complex Objects:

Complex objects are composed of other complex objects or related to them by links. Complex objects do not consist of atomic values only, but can be tuples or sets of tuples of other objects or can have methods as values.

We give below an example of a complex object:

family: father:person;
mother:person;
children:set of person;
person: name:tuple(first:string, last:string);
adr:address;
bdate:date;
age:(currentdate - bdate);
address: street:string;
city:string;
state:string;
date: year:integer;
month:integer;
day:integer;
currentdate:date;

Observe that the slot "age" in the class "person" has a method associated with it. This method is inherited by all instances of the class "person" and invoked when the message "age" is sent to a person instance in order to calculate the age of that person.

The Object Oriented paradigm can be viewed as a refinement of the frame formalism. Frame systems do not distinguish between classes and instances and employ only one kind of relation, the *isa* or *a_kind_of* link. The lack of multiple types of links together with the availability of the cancellation property, which overrides inheritance, can cause logical inconsistencies in the representation schema, (see Brachman, [2]). In object oriented programming, in addition to the *a_kind_of* relationship between classes and subclasses, there is a relationship which holds between classes and instances of those classes, the *an_instance_of* relationship. Another difference between frames and objects is that frames do not employ methods, but only demons which are triggered by some event. A demon can directly access the slot of another frame violating the message passing principle of object oriented systems.

3.5.2 Object oriented paradigm in Prolog (optional)*

In this section, we show how an Object-Oriented programming environment can be built in Prolog. Schema design constraints were added to check for consistent and non-redundant object oriented schema design. An Object Oriented schema is a directed acyclic graph, (DAG). The nodes of the DAG correspond to classes and objects, and the links form an ISA hierarchy. A subclass inherits instance variables and methods from its parent class. Messages are externally sent to an object in the schema and are processed using the inheritance links. Multiple inheritance is encountered when a class is a subclass of many superclasses and inherits properties from them. In this case, we can have the potential for conflicts over which methods should be used or which default value should be inherited. In the implementation given below, the user can resolve multiple inheritance issues, by using depth first search of the DAG (see Chapter 3).

Objects and methods are represented by Prolog facts and rules. A class or object is represented as a collection of Prolog clauses in the following form:

slot_name(Object_name, Value_of_slot)
or
rule_head(Object_name, More_arguments):-body.

For example, the car_1 frame of the previous section is represented under this schema as:

```
sunroof(car_1,yes).
manufacturer(car_1,ford).
lic_plate(car_1,usa346).
basic_price(car_1,10000).
find_price(car_1, P):-sunroof(car_1, yes),basic_price(car_1, B),
                      P is B+500,!.
```

find_price(car_1, P):-basic_price(car_1,P),!.

Objects are linked together via an ISA relationship in the form

isa(Object1, Object2).

Messages are sent using the Prolog query :

send(Object, message, SearchStrategy).

When the system receives this query, it will first try to respond to the message using the methods of the Object which the message is intended for. If this fails, it will follow the ISA relationships toward superclasses, in order to find a parent method that can be applied. The SearchStrategy determines which of the parent methods will be applied in case of multiple inheritance.

New classes and objects are added to the hierarchy by using the predicate

add_object(Super_class, Object, Methods).

Before an object can be added to the hierarchy, schema constraints are automatically being checked. The following three undesirable properties in class hierarchies should be detected:

1. Schema with equivalent classes (cyclic structures).
   If we have the relationships *isa(A,B)* and *isa(B,C)* and *isa(C,A)*, then the classes A,B and C are identical.

2. Schema with redundant isa relationships.
   If we have *isa(A,B)* and *isa(B,C)* and *isa(A,C)*, then the link *isa(A,C)* is redundant, since *isa* is a transitive relationship and *isa(A,C)* does not need to be explicitly stated.

3. Inconsistent schema.
   If we have *isa(A,B)* and *isa(A,C)* and *disjoint(B,C)*, then A must be empty. Here, "disjoint" is a predicate defined as follows:

Classes B and C are *disjoint* iff the intersection of the instances of B and C is the empty set.

The knowledge as to which classes are disjoint has to be provided to the system by the user.

Checking for schema anomalies.
The three schema constraint violations can be detected as follows: First define a set of inference rules

```
isa_chain(A,B):-isa(A,B).
isa_chain(A,C):-isa(A,B),isa_chain(B,C).
disjoint(C,B):-disjoint(A,B),isa(C,A).
```

To detect cyclic structures, define the predicate "inclosure",

```
inclosure(A,X):-isa_chain(X,A), not(isa(X,A)).
cycles(A):-isa_chain(X,A),not(X=A), isa_chain(A,X).
```

The query *?-cycles(A).* will then detect loops. The query, *inclosure(X,a)*, computes the nodes to which there is a path from a of length greater than 1.

To detect redundant ISA links, define the predicate

```
eliminateRedundant:-inclosure(X,Y),isa(Y,X),retract(isa(Y,X)).
```

Then, the query *?eliminateRedundant.* would locate redundant links and retract them from the database.

To check for inconsistency, we define the predicate

```
inconsistent(A):-isa_chain(A,B),isa_chain(A,C),disjoint(B,C).
inconsistent(C):-isa_chain(C,A),disjoint(C,A).
```

The query *?-inconsistent(C).* will respond false, if no inconsistencies exist. If inconsistencies exist, it will respond by giving all the nodes where an inconsistency occurs.

We give below the Prolog predicates implementing the previously defined functions. The given code extends and generalizes the code given in the frames section, (Stabler, [25]).

The *add_object* predicate, which inserts objects into the hierarchy, can be defined as follows:

```
add_object(Superclass, Object, Method):-
        add_method(Object, Method), not(isa(Object, Superclass)),
        assert(isa(Object, Superclass).

add_method(Object, Method):-
        Method=[(Head:Body)], Head=..[Predicate/Arguments],
        Prolog_head=..[Predicate,Object|Arguments],
        assert(Prolog_head:-Body).

add_method(Object, Slot_and_value):-
        Slot_and_value=..[Predicate| Arguments],
        Slot=..[Predicate, Object | Arguments], assert(Slot).
```

The *send* predicate, which sends a message to a class or object in order to query it, can be defined as follows:

```
send(Object, Message):-
        Message=..[Predicate|Arguments], isa_chain(Object, Object1),
        Goal=..[Predicate,Object1|Arguments], (clause(Goal,Body),
        call(Body));true.
```

The predicate create_root is called to create the root of the object oriented hierarchy,

```
create_root:-add_method(root,[description('the root of hierarchy')]).
```

3.5.3 An application

We now look at a subset of the university personnel database, which was created using the predicates given above. The objects are EMPLOYEEs. We assume the following:

a) Employees are classified in one of two types, salaried employee and hourly employee.
b) Salaried employees are Professors, Instructors and Administrators.
c) Hourly employees are Clerical, Maintenance and Graduate Assistants.
d)This is a subset of a larger database schema, which would have additional objects (i.e. Students, Courses, Publications, ...).

The design and construction of an Object Oriented schema involves the following steps:

1. Specify a collection of classes and subclasses, a set of constraints among them, and a set of relationships among them in the hierarchy.

2. Add attributes (or instance variables) and methods to each class and its subclasses. These are the class descriptions and tell why each class is unique or distinguished from other classes.

3. Identify the overlap, (redundant), attributes and methods from all classes. Starting at the top of the hierarchy, eliminate those attributes that are redundant with its subclasses, leaving the attributes at the lowest possible level of subclass. As for the methods, we must start at the lowest level and work our way up, to eliminate those at lower level and leave the ones at the highest level possible.

4. For each declared constraint, an appropriate verification should be performed.

5. Design needed methods based on the type of queries that need to be issued against the data.

A sample OO schema is given in Figure 3.8.

Before adding objects into the schema, the query *"createRoot"* should be issued to create the root ROOT of the hierarchy. To add objects into the hierarchy the "addObject" command is invoked,

add_object(Superclass, Objectname, Objectmethods).

After all objects have been created and the hierarchy built using the ISA predicate, we can issue queries to the database using

send(Object, Message,'df' or 'bf').

The third argument is assigned the value 'df' or 'bf', depending on whether we wish to perform a depth first or breadth first search to resolve multiple inheritance issues.

We give below a sample session, where the university knowledge base is created and queried:

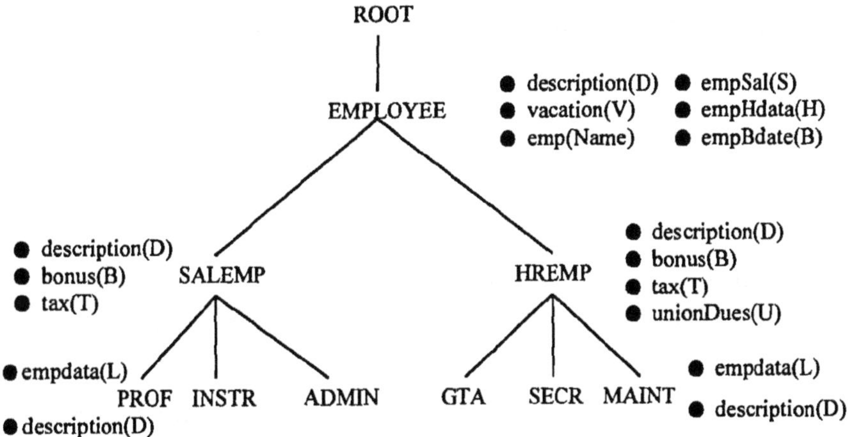

Figure 3.8 The Employee object oriented schema.

?-createRoot.
yes.
/*after execution of above query the following Prolog clauses are generated and asserted:
description(obj,'an object').
index(obj,obj,description('an object')). */

?-add_object(obj,employee(A),
[description('a person who provides work to univ. and paid by univ'),
(empsal(S):-send(A,empdata(L),df), atom-of-list(6,L,S),
(vacation(B):-send(A,empsal(S),df),curyr(CY),B is
*(CY-YS+10)/10*5)]).*
yes.

% after execution of above query following clauses will be asserted in
% the knowledge base:
% isa(employee(P),obj).
% description(employee(A), 'a person who provides work to univ. and
% paid by univ').
%empsal(employee(A),S):-send(A,empdata(L),df),atom-of-list(6,L,YS).
% vacation(employee(A),B):-send(A,empsal(S),df),curyr(CY), B is
% ((CY-YS+10/10)*5.
% index(employee(A),employee,description('a person who provides

```
%work to univ. and paid by univ')).
% index(employee(A),employee,(vacation(B):-send(A,empdata(L),df),
%                                         atom-of-list(6,L,YS),
%                             curyr(CY), B is ((CY-YS+10)/10)*5)).

?-add_object(employee(A),salemp(A),[description('employee has fixed
yearly salary')]).
yes.

?-add_object(employee(A),salemp(A),[(bonus(B):-send(A,empsal(S),df)
, B is S*0.15),(tax(B):-send(A,empsal(S),df),B is S*0.3)]).
yes.

?-add_object(employee(A),hremp(A),[description('employee is paid
hourly'), bonus(B):-send(A,empsal(S),df),B is S*0.15*12),
tax(B):-send(A,empsal(S),df),B is S*0.2)]).
yes.

?-add_object(salemp(prof(81380)),prof(81380),[empdata([smith,tom,
30000, [1934,9,30], cs,1980])]).
yes.

?-add_object(salemp(prof(81234)),prof(81234),[empdata([anderson,
george,39000, [1934,3,2],cs,1980])]).
yes.

?-add_object(salemp(prof(81111)),prof(81111),[empdata([jackson,bo,
49000, [1954,5,7],ms,1984])]).
yes.

?-add_object(hremp(maint(71236)),maint(71236),[empdata([goofy,
arnold,12, [1967,12,15],math,1989])]).
yes.

% We can now issue queries on the created OO schema . For example:
% How many vacation days does professor 81380 have?
?send(prof(81380), vacation(V), df).
V=10 ;
yes.

% Give a listing of all professors with their vacation days
```

```
?send(prof(EmpId), vacation(V), df).
EmpId=81380
V=10;
EmpId=81234
V=10;
EmpId=81111
V=5;
no.

% Get all info about instructor 82222

?send(instr(82222), empdata(Data), df).
Data= [jackson,michael,33000,[1960,8,2],cs,1988]
yes.
```

3.6 PETRI NETS

3.6.1 Definitions

Petri nets were introduced by C. Petri and have been used extensively for analysis of concurrent and asynchronous systems, language simulation, software engineering, communication, process control and knowledge representation. For a detailed introduction to Petri nets, the interested reader is referred to Peterson, ([14]).

A *Petri net* is a directed bipartite graph (there are two types of nodes, and the arcs can only connect nodes of different types). The two types of nodes are called *places* and *transitions*. A place is an *input to a transition* iff there is a directed arc from the place to the transition. A place is an *output to a transition* iff there is an arc from the transition to the place. The in-degree (number of arcs coming into) and the out-degree (number of arcs emanating out of) of a place or a transition can be greater or equal to 0. Places can serve as storage places for *"tokens"*. A transition is enabled and *"fires"*, if all of its input places contain at least one token. Firing a transition involves removing one token from each of the input places and depositing one token into each of the output places of the transition.

A more formal definition of a Petri net is as follows:

A *Petri net* is an ordered 5-tuple <P, T, I, O, m>
where:
P is a finite set of "*places*" $\{p_1,p_2,...,p_k\}$
T is a finite set of *transitions* $\{t_1,t_2,...,t_n\}$
$P \cap T = \emptyset$
$P \cap T \neq \emptyset$
I is the input function from transitions T to 2^P the set of subsets of P.
O is the output function from transitions T to 2^P the set of subsets of P.
A place p_i is an *input place* for transition t_j iff $p_i \in I(t_j)$.
A place p_i is an *output place* for transition t_j iff $p_i \in O(t_j)$.
If $p_i \in I(t_j)$ (or $p_i \in O(t_j)$), then (p_i,t_j) (or (t_j,p_i)) is called an *arc*.

A *marking* m of the Petri net is a vector $(m_1,m_2,..., m_k)$, where m_i is an integer which corresponds to place i and indicates the number of tokens that place i holds ($m_i>=0$).

A transition t *fires* iff for each i with $I(t)=p_i$: $m_i>0$. When a transition t fires, the net marking is adjusted as follows: for each i with $I(t)=p_i$, let $m_i=m_i-1$ and for each j with $O(t)=p_j$, let $m_j=m_j+1$.

A Petri net is initialized with a marking (defining how many tokens each place holds) and transitions start firing. The execution of the net terminates, (the net stabilizes), when there are no more transitions eligible to fire.

Typically, a Petri net is represented by a graph. The place nodes are represented by circles and the places by bars. Tokens can be represented by a number inside a place circle or by a smaller circle. For example consider the net of Figure 3.9, where the places are labeled as a,b and c and the transitions as 1 and 2. When one or the other or both of the places a, b holds a token, then transition 1 or 2 will fire and node c will acquire a token. If we think of a,b,c as corresponding to logic predicates and the presence of at least one token in them as representing the truth value "true", then the net of Figure 3.9 can be thought of as representing the logical or, $c=a \vee b$.

In the following, we describe a Petri net model which can be used to represent knowledge and which has the same expressive power as first order predicate calculus. The model is a variation of the Predicate/Transition model introduced by Genrich, ([8]). The inference mechanism associated with the Petri net representation of knowledge consists of transition firing.

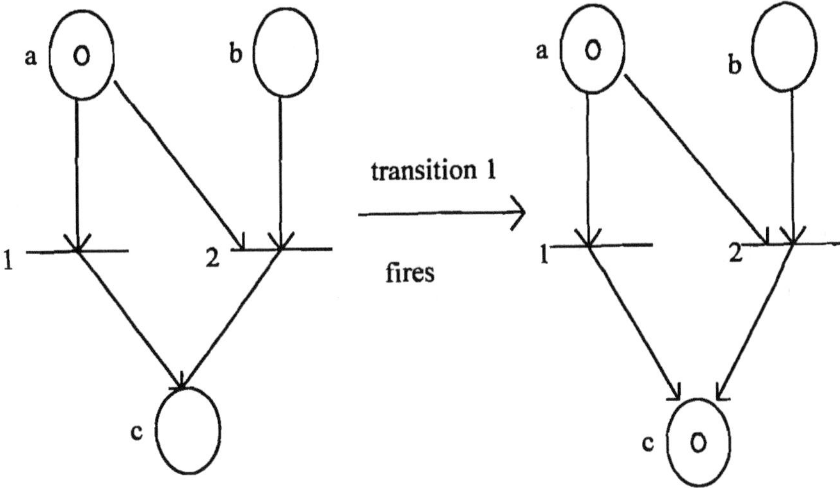

Figure 3.9 A Petri net graph.

3.6.2 Petri net model for propositional calculus

In order to handle logic negation, the basic Petri net model is modified as follows:

1) There are two types of tokens (black and white).
2) Only a white token can traverse an arc from its source to its destination.
3) A black token can only traverse an arc from its destination to the source.
4) A transition can change the type of an arriving token from black to white or from white to black.
5) When all the places which are input to a transition hold at least one token, and the transition is enabled and fires, the tokens at the source places remain in their places and are not deleted.
6) When a transition fires, it deposits a token to at least one of its output places.

The semantic interpretation of the modified model is that *places* correspond to predicates. A white (black) token in a place indicates that the corresponding predicate is true (false). The second type of token was introduced in order to handle negation. Blacks tokens are allowed to travel backwards on the arcs in order to simulate inferencing. For example, the Petri net of Figure 3.10 (a) corresponds to the logic formula $p \Rightarrow q$. When a white token is deposited in place p (p is true), the transition fires, the white token reaches the transition, remains white, and traverses the second arc to reach place q (q becomes true). On the other hand, if a black token is placed in place q, (q is false), the transition is enabled again as all its outgoing places contain a black token, and fires. The black token traverses the first arc backwards and reaches the transition. It can then traverse the arc emanating at p backwards and is deposited to place p. So, the assumption that q is false gives the conclusion that p is false. As a result, the given net is equivalent to the implication $p \Rightarrow q$.

As another example, consider the modified Petri net of Figure 3.10 (b). If a white token is placed in node p, (p is true), the transition is

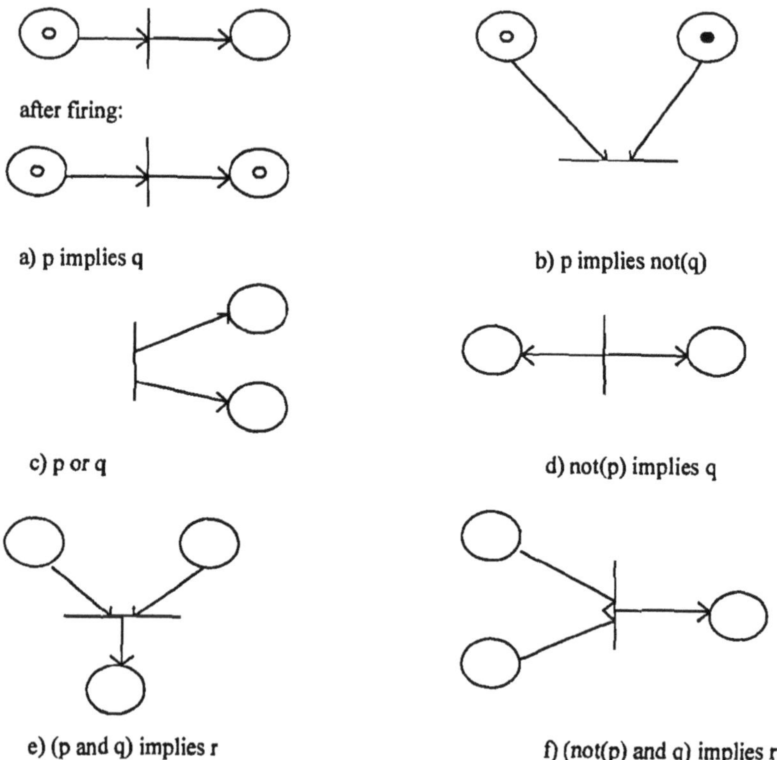
after firing:
a) p implies q
b) p implies not(q)
c) p or q
d) not(p) implies q
e) (p and q) implies r
f) (not(p) and q) implies r

Figure 3.10 Petri net representations of propositional formulas.

enabled. It fires, the white token reaches the transition through the forward arc, where it is changed to black and traverses the second arc backwards to be deposited as black in node q (q is false). Similarly, if a white token is placed in node q, (q is true), the transition will fire and a black token will be deposited in p (p is false). Therefore the net of Figure 3.10(b) simulates the logic implication $p \Rightarrow \neg q$. Consider what will happen if the Petri net corresponding to $p \Rightarrow \neg q$ is given the initial marking: *(white, white)*. The places p and q will be able to deposit black tokens to each other, resulting in places p an q, each holding both a black and a white token. Giving the marking *(white, white)* to the net is equivalent to the knowledge base $\{p \Rightarrow \neg q, p, q\}$. This knowledge base is inconsistent. In general, the presence of both a white and black token in a place node indicates that the knowledge base represented by the net is inconsistent.

Figure 3.10, gives the Petri net representations of some additional formulas.

3.6.3 Petri net model for predicate calculus

In the case of predicate calculus formulas, the existential quantifiers have to be eliminated using Skolem functions before the formula is represented as a Petri net. The propositional calculus Petri net model of the previous section can be modified by allowing the input or output arcs to a place (predicate) to be labeled by the arguments of the predicate. The tokens held by a place are not just boolean values, and a place can hold more than one token. Each token is a record of as many fields as the number of arguments of the place predicate. Tokens are still colored black (white) to indicate whether they make the predicate false (true). A transition is enabled if its input places hold tokens that agree on the arguments shared by the tokens.

As an example consider the following knowledge base:

isa_chain(X,Y):-isa(X,Y).
isa_chain(X,Y):-isa(X,Z), isa_chain(Z,Y).
isa(cat, feline).
isa(feline, animal).

This knowledge base can be represented by the Petri net shown in Figure 3.11.

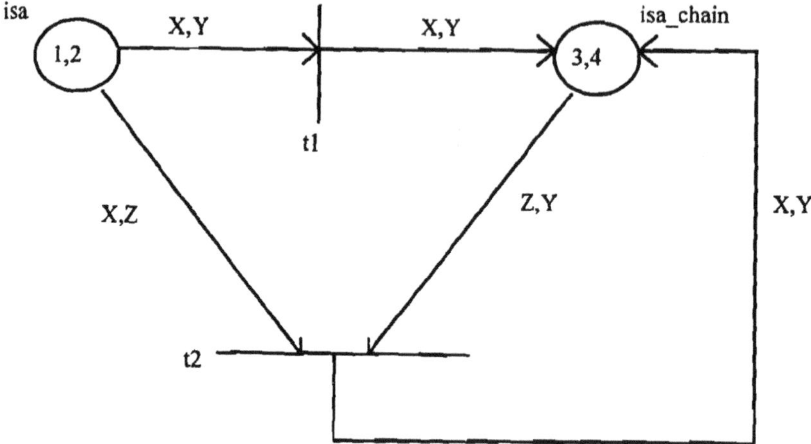

Figure 3.11 Petri net for the transitive closure predicate isa_chain.

The two facts *isa(cat,feline)* and *isa(feline,animal)* are deposited as

token1: (white,cat,feline)

and

token2:(white,feline,animal)

in the place node *isa*. Transition t1 is enabled and fires, passing a copy of token2 into the place node *isa_chain*. Let *token3:(white,feline,animal)* be the new token in place node *isa_chain*. Tokens 2 and 3 cannot enable transaction t2, since they each contain a different value (animal and feline respectively) in the argument Z, which is shared between the two arcs. Transition t2, on the other hand, can be enabled by tokens 1 and 3, which arrive at the transition and combine into *token4:(white,cat,animal)*. Token 4 is then transmitted through the arc (t2, isa_chain) labeled <X,Y> into the place *isa_chain*, deducing that cat is an animal.

3.7 HYBRID REPRESENTATIONS

First generation expert systems typically used one knowledge representation formalism for uniformly representing knowledge across the system, with the rule-based knowledge representation being the most commonly used. Even though this proved adequate for building small to medium scale systems, as developers moved to very large and more complex systems, the rule-based representation did not scale well and major difficulties were encountered during development. Second generation expert systems typically use hybrid representations (combination of multiple knowledge representations) which addresses the fact that different sections of the knowledge base may be encoded more efficiently and easily using different formalisms.

A hybrid knowledge representation which is available in most current expert system shells today combines rules, frame based and object oriented techniques, facilitating the model based approach to system development and making the development of complex models and class hierarchies much easier than when using the rule based approach by itself. Several commercial expert system shells, including ADS, ART, KEE, Knowledge Craft and ProKappa, (see appendix C), employ an object oriented based hybrid representation scheme. They supply a user friendly definition language for creating classes and objects. Object instances can be created and stored in databases and retrieved when needed through built in interfaces. The presence of the OO environment also makes it easier to develop customized user interfaces. The rule based component consists of *pattern-matching rules* which are associated with the objects. A pattern matching rule performs a function similar to the join operator in relational databases. Its conditions check if the attributes of objects have a certain value or compare the values of attributes of various objects. Their action part can send a method to an object as a message to create object instances or alter attribute values of an object.

As an example, consider the employee class hierarchy of Section 3.5.3 and the classes PROF (Professor) and GTA (Graduate Teaching Assistant) with the additional attributes "specialty" and "interest" respectively used to indicate their areas of expertise and interest. In the Prolog based OO environment of that section, a pattern matching rule for assigning graduate teaching assistants to professors based on common interests would be:

assign(P, G):-send(prof(P), status(T1),df), send(gta(G), status(T2), df),
T1=not_matched, T2=not_matched, send(prof(P),
specialty(S), df), send(gta(G), interest(I), df), S=I,
send(prof(P), match(matched), df), send(gta(G),
match(matched), df).

In the commercial expert system development environment ADS, (Aion Development System), the same pattern matching rule for assigning graduate teaching assistants to professors, augmented by giving priority to younger professors, would have the form:

IF PROF?specialty=GTA?interest
and GTA?status=not-matched
and PROF?status=not-matched
ORDERBY(PROF.empBdate)
THEN create(pairs with PROF.empName, GTA.empName)
and PROF?match
and GTA?match.

where "match" is a method attached to the PROF and GTA classes, which changes the value of the "status" attribute to matched, and ORDERBY is a built in method attached to the root class which performs hypothetical reasoning by using the A* algorithm to rank solutions. The A* algorithm orders instances of objects using a specified attribute (or a combination of attributes) and pattern matches rule conditions with object instances in that order.

Pattern matching rules provide a convenient way which is employed by most object oriented/rules hybrid systems for communicating and passing knowledge between the two different paradigms. In general, a difficulty with using multiple representations is the need to pass knowledge between the different paradigms. An interesting approach suggested by researchers is to develop a generic representation language and procedures for translating knowledge from any other formalism to this canonical form. This way, there is no need to develop procedures for translating knowledge between all possible pairs of representation schemes, but we only need to implement procedures to translate a representation to the canonical form and back (Simmons, [22], Rowley et al., [17]).

3.8 THE FRAME PROBLEM

The *frame problem* is defined as the problem of finding a representational form permitting a changing, complex world to be efficiently and adequately represented, (for an analysis of the frame problem see [9]). When a change occurs in the world, the majority of things are not effected by it. How can we know which are the blocks of knowledge affected by it and which need to be updated without exhaustively checking all of the knowledge? Trying by brute force to predict all possible correlations in advance and include them in the knowledge base, clearly won't do in complex domains, as it introduces combinatorial explosion. The frame problem concerns a truly dynamic real world environment, in which circumstances that could be encountered are too vast to be modeled in a static manner. Present knowledge representation techniques involve the modeling of an idealized subset of a real world problem and suffer from their inability to adjust in unforeseen circumstances not predicted by the model. No matter which representation we choose, its effectiveness is constrained by the frame problem.

3.9 SEMANTIC PRIMITIVES

Another problem relevant to knowledge representation is the issue of establishing a representational vocabulary which maintains uniqueness of knowledge representation structures. If knowledge is available in natural language form, the same conceptual meaning can be expressed in many different ways. For example, *I received a book in the mail* or *I got a book which was sent to me by mail* conceptually represent the same knowledge, but they could be expressed in a representation, such as logic or semantic nets, in more than one way. It is not desirable to have multiple representation structures in a knowledge base represent the same underlying knowledge. This can lead to inefficiency and confusion. As a solution, we should develop techniques to determine when two representations are equivalent (a very difficult problem in the general setting). When duplication is detected one of the two equivalent representations should be eliminated or the knowledge engineer should be warned about the existence of duplication. In another approach, a set of semantic primitives has been proposed, (for example, in Schank's conceptual dependency and scripts theory, [19]), so that every representation can be expressed uniquely using these primitives as building blocks.

SUMMARY

Knowledge representation is a very important step in building intelligent and expert systems. It encodes the knowledge acquired through knowledge acquisition in a form suitable to computer manipulation. There are no developed formal methodologies for deciding which knowledge representation technique to choose for a given problem, but a good choice will go a long way in facilitating the solution process. This chapter introduced several well known representation techniques including production rules, semantic nets, frames, the object oriented formalism and Petri nets. Large, complex systems usually employ hybrid representations, where more than one representation formalisms are combined in the same system to efficiently represent different sections of the knowledge base.

EXERCISES

1. Design a semantic network for the organizational structure of your college, department or company, including human resources such as faculty members or employees.

2. Rewrite the semantic net of Exercise 1, as a conceptual graph with actors that calculate the total number of faculty (or employees) in each department.

3. Describe the prerequisite structure of courses in your major field of study using the production rule representation.

4. Rewrite the rules in Exercise 3 in such a way that given any course as a query, the system gives all the prerequisite courses that have to be taken before this course. Implement the rule base you created using the expert system shell available to you.

5. Modify the rules of Exercise 4, so that when a list of all the courses taken so far is given, the system returns a schedule of the core courses that remain to be taken in prerequisite order. Implement the rule base using the expert system shell available to you.

6. a) Draw a Frame based representation for the college/department structure at your University, with if-needed methods for calculating the total number of faculty per department college and the whole university.
   b) If you have access to an expert system shell which supports the frame formalism (like for example EXSYS) implement the same frame system using the shell.
   c) Implement the frame system and a query machine in Prolog.

7. If the expert system shell available to you supports the Object Oriented paradigm (for example LEVEL5-Object or JESS, implement an object oriented schemata for the knowledge base in Exercise 6 using the shell.

8. a) Write a Prolog program to trace a semantic net and answer queries.
   b) Use your program as an inference engine for Exercise 1.

9. Repeat Exercises 4, 5, 6 and 7 using Prolog instead of a shell.

10. Write the following Prolog procedures to extend the object oriented environment of Section 5.6.

    a) objectname(Objectname) returns the objects in a knowledge base.
    b) methods(Object, M) returns the methods associated to object *Object*.
    c) unlink(Object1, Object2) removes the isa link between the two objects.

11. Extend the Prolog implementation of Object Oriented programming, so that it supports multiple inheritance. (Use either depth first search or breadth first search to choose which parent to inherit from).

REFERENCES

[1] Bobrow, D.G., Winograd, T., "An overview of KRL, a Knowledge Representation Language", *Cognitive Science*, 1,3, 1977.

[2] Brachman, R.J. and Levesque, H.J., *Readings in Knowledge Representation*, Los Altos, CA, Morgan Kaufmann, 1985.

[3] Brachman, R. J., I lied about the trees, *AI Magazine*, Vol. 6, (3), 1985, pp. 80-93, .

[4] Dechter, R., Meiri, I., Pearl, J., "Temporal Constraint Networks", *Artificial Intelligence*, 49, 1-3, 1991, pp. 61-96.

[5] Etherington, D., Reiter, R., "On Inheritance Hierarchies with

Exceptions", *Proceedings National Conf. A.I.*, MIT Press, Cambridge, MA, 1983, pp. 104-108.

[6] Forbus, K., Nielson, P., Faltings, B., "Qualitative spatial reasoning: the CLOCK project", *Artificial Intelligence*, 1, 1-3, 1991, pp. 417-471.

[7] Forgy, "RETE: A fast algorithm for the many pattern, many object match problem", *Artificial Intelligence*, 19, 1, pp. 17-37.

[8] Genrich, H.J., "Predicate/Transition Nets", *Lecture Notes in Computer Science*, Springer Verlag, 254, 1987, pp. 207-247.

[9] Janlert, L.E. "Modelling change-the frame problem". In: *Pylyshyn, Z.W. The Robot's dilemma, the frame problem in Artificial Intelligence*, Norwood, NJ: Ablex Publishing, 1987.

[10] Koton, P. "Combining causal models and case-based reasoning", *in: J. David, J. Krivine, R. Simmons, Eds., Second Generation Expert Systems*, Springer Verlag, 1992.

[11] Lim, E., Cherkassky, V., "Semantic Networks and Associative Databases", *IEEE Expert*, August 1992, pp. 31-39.

[12] McCarthy, J., "First Order Theories of Individual Concepts and Propositions", in: *Brachman, R. and Levesque, H. (eds.) Readings in Knowledge Representation*, Morgan Kaufmann, 1985.

[13] Minsky, M., "A Framework for Representing Knowledge", in: Winston, P., (ed.) *The Psychology of Computer Vision*, New York: McGraw Hill, 1979, pp. 211-277.

[14] Peterson, J.L., *Petri Net Theory and the Modelling of Systems*, Prentice Hall, Englewood Cliffs, NJ, 1981.

[15] Pfeiffer, H.D., Hartley, R.T., "The Conceptual Programming Environment", in *Nagle, T. et al. (eds.), Conceptual Structures: Current Research and Practice*, Ellis Horwood, 1992, pp. 87-107.

[16] Quillian, M.R., "World Concepts: A Theory and Simulation of some basic Semantic Capabilities", in: *Behavioral Science*, vol. 12, 1967, pp.410-430.

[17] Rowley, S., Shrobe, H., Lassels, R., "Joshua:Uniform access to heterogeneous knowledge structures", *Proceedings of AAAI-87*, Seattle, WA, 1987.

[18] Schank, R. C., "Conceptual Dependency: A Theory of Natural Language Understanding", *Cognitive Psychology*, Vol. 3, # 4, 1972.

[19] Schank, R. C., *The Cognitive Computer: On Language, Learning and A.I.*, Reading, MA: Addison Wesley, 1984.

[20] Shastri, L., *Semantic Networks: An Evidential Formalization and its Connectionist Realization*, Pitman, London, 1987.

[21] Simmons, R. and Slocum, "Generating English discourse from Semantic Networks", *CACM*, 15, 1972, pp. 891-905.

[22] Simmons, R., "Integrating multiple representations for incremental, causal simulation", *Proceedings Confer. on AI, Simulation and Planning*, 88-96, Cocoa Beach, FL, April 1991.

[23] Simmons, R., "The roles of associational and causal reasoning in problem solving", *Artificial Intelligence*, 53, 2-3, 1992, pp. 159-208.

[24] Sowa, J.F., *Conceptual Structures: Information Processing in Mind and Machine*, Reading, MA: Addison Wesley, 1984.

[25] Stabler, E. "Object Oriented Programming in Prolog", *AI Expert*, October 1986, pp. 46-54.

[26] Touretzky, D.S., *The Mathematics of Inheritance Systems*, Pitman, London, 1986.

[27] Winston, P., *Artificial Intelligence*, Reading, MA: Addison Wesley, 1987.

CHAPTER 4

KNOWLEDGE ACQUISITION, VERIFICATION AND VALIDATION

4.1 THE EXPERT SYSTEM DEVELOPMENT PROCESS

The term *Knowledge Engineering* is used to describe the process of expert system development (Feigenbaum, [17]). The individual responsible for the expert system development is called a *knowledge engineer.* Knowledge engineers elicit the expert's domain knowledge, model it in a suitable representation scheme, implement the system and refine it with the help of the domain experts until satisfactory performance is achieved. In first generation expert systems, knowledge engineering was viewed as simply the process of extracting knowledge from the human expert and representing it in a form amenable to computation. While this is certainly true, second generation expert systems view the knowledge engineering process more as a modeling activity, in which the knowledge engineer constructs a knowledge level model of the system and instantiates the model with the specific domain knowledge. The knowledge engineer can design a knowledge level model for the application or can choose the model that fits the application from a set of predefined models. The construction of the model and its instantiation with domain specific knowledge are not necessarily sequential processes, but may interleave. For example, during domain knowledge acquisition it may become evident that the chosen knowledge level model is inadequate and needs to be modified.

Typically, the expert system development process starts by developing an understanding of the problem domain and the major concepts involved. The problem and the system goals have to be identified. The intended uses of the expert system and the targeted users have also to be identified early on. These tasks correspond to the system analysis phase of the traditional software engineering life cycle approach.

System design entails the processes of knowledge acquisition (elicitation of knowledge, modeling the domain, fitting the domain specific knowledge in the model, expressing the knowledge structures in an appropriate knowledge representation form), user interface design and the selection of software and hardware.

Implementation involves choosing the shell or the programming language environment which is to be used to implement the conceptual model of the elicited knowledge. It also involves the building of the user interface, explanation facility and interfaces to other systems, like databases or spreadsheets.

Testing proceeds in parallel with the phases described above. It takes place during all stages of development and includes verification and validation of parts of the expert system and of the system as a whole.

During expert system development, a *rapid prototyping* approach is often employed, in which a prototype is built very early on, based on a subset of the domain knowledge. The intention of this approach is to demonstrate feasibility and to make a running version of the system available for inspection early on. This can lead to easier knowledge acquisition and expansion of the knowledge base and facilitate communication by providing better understanding of the problem and the issues involved. The prototype does not necessarily have to evolve to the final product. Often it is discarded after it serves its purpose in the system analysis and design stages. For example, a prototype may be built using a shell to achieve speedy development, but the final system may be built in a programming language for efficiency and flexibility reasons.

Knowledge acquisition consists of two components, the *elicitation* of knowledge and the *representation* of knowledge. Many knowledge representation schemes including semantic networks, frames, rules, the object oriented formalism and Petri nets were described in Chapter 3. The knowledge engineer should not determine which representational mechanism or which tool is to be employed before analyzing the problem type and the domain knowledge. If this is not done, the building of the conceptual model can be influenced, as the knowledge engineer may try to fit the knowledge in the pre-chosen representation scheme, regardless of suitability, and be tempted to ignore whole sub-areas of knowledge which cannot efficiently be represented in that scheme.

4.2 KNOWLEDGE ELICITATION

4.2.1 The elicitation process and the bottleneck

In *knowledge elicitation*, the knowledge engineer elicits the knowledge which is necessary for problem solving in the domain of discourse. To do so, the knowledge engineer may use various means at his/her disposal, such as interviewing the domain experts, using automated knowledge elicitation tools, consulting books, manuals, etc. A conceptual model of the domain knowledge is created as the result of the elicitation process. In second generation expert systems, knowledge elicitation is considered more as a domain modeling activity and not just a simple knowledge transfer from the expert to the computer.

In general, creating an accurate model of an expert's knowledge is a major obstacle in expert system development, known as the *knowledge acquisition bottleneck* (Feigenbaum,[17]). Knowledge elicitation is a difficult and time consuming task. It is also a very important task, as the content of the knowledge base is what determines to a large extent the quality of the expert system. Significant progress has been made in the last 20 years and knowledge acquisition tools are now available to the knowledge engineer, which can facilitate the transfer and modeling of the knowledge. Even so, the available shells and tools are still somewhat limited in the variety of tasks which they can provide assistance with, and the areas of knowledge acquisition, validation, verification and maintenance remain as major areas of concern.

4.2.1.1 Procedural versus declarative knowledge

After repeated application of a particular problem solving activity, human experts seem to subconsciously convert their expertise from declarative to procedural knowledge. Declarative knowledge is accessible to consciousness, while procedural knowledge is not (Anderson, [1]). A domain expert usually has difficulty articulating or explaining to the knowledge engineer any procedural tasks which he performs subconsciously, routinely and automatically. In order to provide explanations of his problem solving activities, the expert has to reconstruct the declarative knowledge that he once possessed and which led to his present state of procedural knowledge. This reconstruction may or may not agree with how he actually solves the domain problems and is only an approximation of the original declarative knowledge. As a result, the process of knowledge acquisition is not just a transfer of knowledge from the expert to the engineer to the computer, but in reality a model of the knowledge is created. The expert and the knowledge engineer each start with their individual conceptual models of the problem domain. Knowledge acquisition can be viewed as the process of conversion to a common conceptual model (Regoczei, [37]). Even assuming perfect interpersonal relationships and cooperation between the expert and the knowledge engineer, the problem of miscommunication and bridging the conceptual gap between the expert's and the engineer's knowledge models, can be a serious problem. Terminology used by experts sometimes seems as too vague, open to interpretation and not well defined to the knowledge engineer. The agreement and construction of a "systematic domain" which provides a formal specification of the task and defines the domain terms seems to be helpful in avoiding the miscommunication problem, (Winograd and Flores, [48]).

Another factor which can make the knowledge acquisition process more difficult is the presence of uncertainty in the domain and the expert's problem solving methodologies and decisions. Many of the developed methodologies for dealing with uncertainty in expert systems do not really correspond to the way in which human experts naturally reason. Forcing the experts to quantify uncertainty by providing numerical values, for example in the form of probabilities, and in spite of this not being the way the expert reasons, can cause misrepresentations in the knowledge base.

The remainder of this section contains an introduction to techniques of knowledge elicitation. For a classification and coverage of additional elicitation techniques, the reader is referred to [12].

4.2.2 Interviews

An interview is an elicitation session between one or more engineers and one or more experts. An interview can be *structured* or *unstructured, focused* or *unfocused.* A structured interview consists of a format predefined by the engineer and a set of goals, which are to be pursued, and information to be gathered. In an unfocused or open interview, the knowledge engineer is looking for knowledge in general, while the objective of a focused interview is to address a specific subarea of the problem solving task. Initially interviews tend to be unstructured and open ended and, as the engineer gets more familiar with the domain, they tend to be more structured and focused.

One way of contacting an interview is by *direct questioning.* When doing so, it is not only a matter of how to ask questions, but also a matter of which questions to ask. There are in general two types of questions: the ones which ask for specific information and those which only require consent or disagreement to the question contents. A question can be *open* or *closed*. For example, the expert can be asked to explain how he would decide if a given textile is rayon or not (closed question) or to describe any difficult textile identification cases he recalls (open question). Usually, questions that begin with *why, what, how* are open ended; questions that start with *who, where, when, which* are of the closed type.

Cognitive psychology research suggests that when humans learn, they tend to segment the related information into various segments (Anderson, [1]). This seems to imply that different questioning techniques should be used to most effectively extract pieces of information of the various types. In a paper by LaFrance, ([25]), five different types of knowledge are introduced. Namely, *layouts, stories, scripts, metaphors* and *rules of thump*. A *layout* type of knowledge contains the overall strategy, methodology and justification of why the expert uses the specific domain knowledge the way he does. Knowledge in the form of *stories* represents examples and past cases from the expert's experience. A *script* describes sequential and procedural knowledge, while *metaphors* attempt to understand the unique characteristics of a concept by comparing it to another concept. *Rules of thumb* correspond to problem solving heuristics.

Based on this classification, five different types of questions can be introduced, each specializing in certain knowledge types. They are: *grand tour questions, cataloging the categories, ascertaining the attributes, determining the interconnections, seeking advice and cross checking questions*.

In the *grand tour* type of question, the expert is asked to give a broad overview of the problem and the domain with the main intention of eliciting the layout type of knowledge. A *cataloging the categories* type of question attempts to discover any hierarchical structure or subsumptions among the domain concepts. *Ascertaining the attributes* types of questions are aimed at finding out distinguishing characteristics of a concept and acquires knowledge in the form of stories or metaphors. *Determining the interconnections* attempts to uncover relationships among concepts and usually addresses scripts type of knowledge. *Cross checking* type of questioning intends to further clarify or validate the information obtained up to that point. Examples of such questioning include "playing dumb", "posing a hypothetical question", etc.

Direct questioning has certain drawbacks. First of all it is difficult to find the right questions to ask. It is more complicated than asking *why* or *how*. The answer often seems to depend on the way questions are phrased. Indeed, research in Psychology has shown that different types of questions elicit different types of answers (Clark, [10]). A large repertoire of various questioning techniques can be employed to make interviewing more effective and optimize the information received. Besides direct questioning, the engineer can ask the expert for guidance in performing the task at hand or ask the expert to reason backwards from a classification, goal or solution to what evidence is required to derive that goal. Most of the questioning techniques were adopted from psychology. A knowledge engineer who is aware of the various questioning techniques and uses appropriate questioning strategies will be more successful at the knowledge elicitation task. For a detailed reference to various interview techniques, the reader is referred to Scott, Clayton and Gibson, ([41]). For a more detailed exposition of questioning methodologies, the reader is referred to Hoffman, [23].

Other elicitation techniques, which are designed to complement direct questioning and avoid some of its pitfalls include repertory grid techniques and protocol analysis. They are discussed in the following sections.

4.2.3 Repertory Grids

The *Repertory Grid* technique is based on the Personal Construct Theory and is designed to answer *ascertaining the attributes* types of questions. It attempts to discover the distinguishing features of the concepts which the human expert is using in his problem solving activities.

Personal Construct Theory was introduced in Psychology by G. Kelly, ([24]), and the process was first automated by Shaw, ([44]). It has been extensively used in clinical psychology and successfully adopted as a knowledge acquisition technique in expert system development. In Personal Construct Theory, an *element* is a concrete or abstract entity, which constitutes an important and relevant case from an expert's personal perspective. A *personal construct* is an important attribute, (according to the judgement of the problem solver), which can be used to distinguish the elements from each other or verify their similarity. The constructs are represented by pairs of *bipolar concepts*, with the relationship between the two pole concepts being one of contrariety, like happy-sad. During the knowledge acquisition process, the knowledge engineer asks the expert to provide all of the elements and the personal constructs which distinguish them.

The elicited knowledge can be stored in a repertory grid. A *repertory grid* is a table whose rows are bipolar constructs and the columns correspond to elements. The set of elements used in the grid constitute the universe of discourse of the grid. Given a bipolar construct C_i and an element E_j, the (i,j)-th entry of the repertory grid is a value indicating the closeness of the element to one or the other of the bipolar concepts in the construct. Each element can then be represented by a vector whose entries are the element's values for the various constructs. Once constructed, the grid can be analyzed to provide further insight to the expert and the engineer and to derive information about associations and dependencies of constructs.

The relationship of an element to the pole of a bipolar concept can be described by the value of 1 or 0 (1, if the element is described by one pole concept of the construct and 0 if it is described by the opposite pole). For example, given the bipolar construct: C_i =happy-sad, and the element: E_j=Christian, then the (i,j)-th element of the grid matrix, grid(i,j), is defined as 1, if Christian is happy, and 0, if Christian is sad.

The range of ratings used to describe how an element fits into a construct can be extended to a discrete set, for example a value from 1 to 5. With respect to the construct *happy-sad*, a rating of 1 would then represent an element which is extremely happy, 5 an element which is extremely sad. An individual receives a ranking of 2, if he/she is somewhat happy (more happy than sad), and a rating of 4 if he/she is somewhat sad (more sad than happy). A rating of 3 is the midpoint and can be construed to express either neutrality (neither happy nor sad), or irrelevance (the element is a robot, so it has no emotions).

Table 4.1 presents a subsection of a repertory grid which appeared in Gaines and Shaw, ([19]), and Ford et al., ([20]). The elements of the domain of discourse are company staff members and the objective of the grid development was to aid knowledge acquisition of the staff appraisal process. The bipolar concepts playing a role in appraisal as well as individual ratings for the staff members were elicited from supervisors.

Table 4.1 A Sample Repertory Grid.

	ELEMENTS					
	Mike	Joe	David	Jill	Mary	
	1	2	3	4	5	
intelligent	1	1	4	5	3	bonehead
introspective	1	1	5	4	2	superficial
beginner	1	2	3	5	4	experienced
motivated	1	1	4	5	2	unmotivated
reliable	3	2	2	5	1	undependable

Several approaches to grid analysis have been proposed, including factor analysis, multidimensional scaling, cluster analysis and fuzzy set theory. PLANET, ([45]), is one of the software tools developed for grid elicitation and cluster based analysis. ENTAIL, (Gaines and Shaw, [19]), is an algorithm which analyzes repertory grids based on fuzzy set theory and each construct is viewed as a fuzzy set. Fuzzy set theory is used to derive *entailments* (rules of the form "Every A is B") from the grid. ETS is a tool for elicitation and analysis of grids using ENTAIL, (Boose, [4]).

Some weaknesses inherent in the repertory grid technique have been cited in the literature, (Batty and Kamel, [2], Rugg and Shadbolt, [39]). For example, the constructs elicited from the domain expert are totally subjective. As a result, validation becomes difficult. It cannot easily be ensured that the choices made for elements and constructs truly represent the given problem. Another concern is that eliciting and analyzing numerical ratings from the experts does not necessarily correspond to the way experts think. As a result, the ranking given to an element with respect to a construct is not necessarily a measure of the element's relevance to one or the other concept poles of the construct. Midpoints in the scale are particularly difficult to assign a meaning to. Do they represent neutrality or irrelevance? For an alternative to the repertory grid technique see the paper by Batty and Kamel, ([2]). Their approach is still based on the Personal Construct Theory, but the knowledge elicitation is based on a network structure to address some of the weaknesses mentioned above.

The next two sections describe two of the methods used for the analysis of repertory grids. The first is based on the theory of confirmation and the second on cluster analysis.

4.2.3.1 Repertory grid analysis using NICOD

NICOD is a automated knowledge acquisition system which uses the theory of confirmation for analyzing a repertory grid and producing entailments between concepts of the form *Every A is a B* (Ford et al., [18]). Its basic algorithm for rule generation is briefly described below, by using the grid of Table 4.1 as an example.

The ratings of the elements in the domain of discourse of the grid are converted from the scale 1,2,3,4,5 to 1.0, 0.75, 0.50, 0.25, 0.0 respectively. The repertory grid is converted to a set of α-planes, where α=1.0, 0.75, 0.5, 0.25 or 0.0. An *α-plane* is a matrix with rows labeled by the constructs and columns labeled by the elements in the grid's domain of discourse. If the rating of the relationship of element j to concept i is α, then the (i,j)-th entry of the α-plane matrix is equal to 1, otherwise it is equal to 0. See Table 4.2 for the set of α-planes equivalent to the repertory grid of Table 4.1.

Table 4.2 α-planes for the Staff Members Repertory Grid.

α's	1.0	0.75	0.50	0.25	0.0
individuals:	12345	12345	12345	12345	12345
intelligent	11000	00000	00001	00100	00010
introspective	11000	00001	00000	00010	00100
beginner	10000	01000	00100	00000	00001
motivated	11000	00001	00000	00100	00010
reliable	00001	01100	10000	00000	00010

The concepts at the two opposite poles of a construct are labeled as "extreme" at the 1.0 and 0.0-planes, "somewhat" at the 0.75 and 0.25-planes and the concepts at the 0.50 plane are neutral relationships. For example, the new concepts "extremely intelligent", "somewhat intelligent", "extremely bonehead" and "somewhat bonehead" are created.

Each α-plane is used to derive entailment relationships of the form "every A is B". Consider for example the 0.0-plane, and the concepts "extremely unmotivated" and "extremely unreliable". These concepts are represented by the vectors 00010 and 00010, respectively. Consider the entailment: *extremely_unmotivated ⊐ extremely_unreliable*. The *total confirmatory evidence* in support of such an entailment is defined as the number of positions where both vectors have the value 1 (i.e. number of elements which are extremely unmotivated and unreliable). The *total relevant evidence* is the number of elements which have the property *extremely unmotivated* (i.e. the number of ones in the vector "extremely unmotivated"). A measure of the *degree of confirmation* for the entailment is given as the ratio of the total confirmatory evidence divided by the relevant evidence. In this example, the strength of the generated entailment is 1/1=1.0.

As another example, consider the entailment *extremely introspective ⊐ extremely beginner* at the 1.0-plane (corresponding vectors 11000 and 10000). The number of elements which are both introspective and beginners is 1, while two elements are introspective. Therefore, the strength of the entailment *extremely_introspective ⊐ very_much_beginner* is 0.50.

4.2.3.2 Repertory grid analysis using clustering

Consider a set of n objects and m attributes, which are used to characterize the objects. Each object is represented as a vector of m values, one for each attribute. In cluster analysis, a *similarity measure* is defined as a function which associates a numerical value to any pair of vectors to indicate how similar the vectors are. The *proximity matrix* is an n by n diagonal matrix with the (i,j)-th entry equal to the proximity(similarity) measure of objects i and j. The proximity matrix is the input to a clustering algorithm. The objective of a clustering algorithm is to form groups (clusters) of similar elements. Various clustering algorithms have been proposed including hierarchical clustering and k-means clustering. For a thorough introduction to statistical cluster analysis see Dubes, ([14]). Various techniques from statistical cluster analysis have been transferred and applied to the knowledge elicitation area. For a detailed description of such techniques, the reader is referred to Myer et al., ([32]). In this section, we give a brief exposition of how hierarchical clustering can be applied to focus a grid.

In hierarchical clustering, the proximity matrix is transformed into a sequence of nested partitions of the set of objects. A *clustering* is a subdivision of the set of objects, E, into a set of subsets, E_i, covering E, and such that $E=\cup_i E_i$ and $E_i \cap E_j = \varnothing$, for $i \neq j$. The sequence of clusterings created by hierarchical clustering depends on the proximities of elements only through their rank order. There are a number of functions that can be used to measure proximity of two vectors, $u=(x_1,x_2,\ldots,x_n)$ and $w=(y_1,y_2,\ldots,y_n)$, such as the taxi-cab metric, $d(u,w)=\sum |x_i-y_i|$, or the euclidean distance of the vectors, $d(u,w)=\sum |x_i-y_i|^2$. Both of these measures are dissimilarity measures, in the sense that the more dissimilar the vectors are, the greater the measure. They can be converted to a similarity measure, (the more similar the objects, the greater the measure), by $s(u,w)= [\sum D_i-d(u,w)]/\sum D_i$, where D_i is equal to the absolute value of the difference between the maximum possible value and the minimum possible value of the ith coordinate (the domain range). The measure is divided by $\sum D_i$, the maximum possible distance between two vectors, for normalization purposes.

Hierarchical clustering can be used to cluster both constructs and elements in the domain of discourse. If the repertory grid has m constructs (m rows) and n elements (n columns), then each construct corresponds to a vector of dimension n, whose ith coordinate is the construct rating of the ith element. Each element corresponds to a vector of dimension m, whose ith coordinate is the rating of the element for the ith construct. In Shaw's FOCUS algorithm, ([42]), clustering analysis techniques are used to focus the repertory grid, so that similar elements and constructs are close to each other in the grid.

In the case of the repertory grid of Table 4.1, all ratings are in the range 1 to 5, so $D_i=5-1=4$, for each i. The similarity measure between two vectors becomes $(4*n-\sum|x_i-y_i|)/(4*n)$. When two vectors are completely opposite, ($|x_i-y_i|=4$, for all i), then their similarity measure is 0, and if they are identical then their similarity is 1. Consider for example, the vectors for Mike and Joe in the staff evaluation repertory grid. Their similarity is measured as:

similarity([11113],[11212])=(4*5-2)/(4*5)=18/20=0.90.

For the grid of Table 4.1, the similarity matrix between elements is given in Table 4.3.

Table 4.3 Similarity Matrix for Elements in the Staff Evaluation Repertory Grid.

		Mike	Joe	David	Jill	Mary
		1	2	3	4	5
Mike	1	1	0.9	0.35	0.15	0.55
Joe	2		1	0.45	0.15	0.65
David	3			1	0.60	0.40
Jill	4				1	0.60
Mary	5					1

Hierarchical clustering is started by joining into a cluster the two elements of the greatest similarity. It proceeds by traversing the similarity matrix in descending order of its entries and joining the corresponding elements into clusters. If an element already belongs to a cluster, the element joining it also becomes part of the cluster. The process can be depicted by a dendrogram as in Figure 4.1. At the end of the procedure all elements are joined into one cluster. Horizontal cross-sections of the dendrogram at different levels give rise to different possible clusterings.

Constructs can be clustered in a similar fashion, but similarities have to be evaluated not only between constructs, but also between constructs and their bipolar opposites. For example, a construct such as short-tall may not be similar to the construct heavy-light, but it may be related to the opposite construct light-heavy. The similarity matrix between constructs and opposites is shown in Table 4.4 and the dendrogram in Figure 4.2. In Table 4.4, the matrix entry at coordinate (i,j) indicates the similarity measure between construct i and construct j, if i<j. For i>j, the value in the (i,j)-th entry denotes the similarity between the bipolar construct of construct j and construct i. For example, the (3,1) entry is equal to 0.35, which is the similarity between the construct *bonehead-intelligent* and *beginner-experienced.*

Based on the clusterings, the repertory grid can be rearranged so

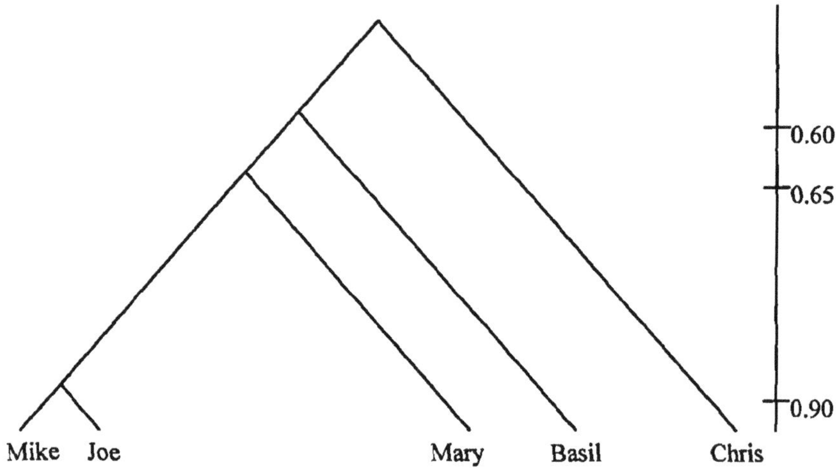

Figure 4.1 Dendrogram for hierarchical clustering of staff evaluation grid.

that similar concepts and elements are close together. This results in a *focused grid.* The domain expert may be prompted to update his views on which constructs or elements are important for inclusion and which are redundant, or may become aware of new relationships between constructs or elements. Derivation of construct entailments is also possible by using a focused grid.

Table 4.4 Similarity Matrix between Constructs in the Staff Evaluation Repertory Grid.

	Constructs				
	intelligent	introspec-tive	beginner	motivated	reliable
intelligent	1	0.85	0.85	0.95	0.70
introspec-tive	0.25	1	0.70	0.90	0.60
beginner	0.35	0.40	1	0.80	0.70
motivated	0.25	0.20	0.40	1	0.70
reliable	0.45	0.40	0.50	0.40	1

The corresponding hierarchical clustering dendrogram is given in Figure 4.2.

4.2.4 Protocol analysis

Protocol analysis is a structured technique for extracting knowledge from a text knowledge base. During the interview process the knowledge engineer often asks the experts to solve a problem in their domain and think out aloud as they are solving it, reporting the subgoals they prove and the evidence which guided them to reach these subgoals. The elicitation session can be taped and transferred to text form. Using protocol analysis, the text can be analyzed by extracting from it three basic categories of behavior: *operations, episodes and operands*. An *operation* is usually indicated in the text by a verb and is a generic task like measuring the water temperature or performing a computation. An *episode* is a sequence of operations repeatedly used during problem solving activity, while *operands* are data manipulated by the operations. For a detailed coverage of the protocol analysis technique, the interested reader is referred to Ericsson and Simon, ([16]).

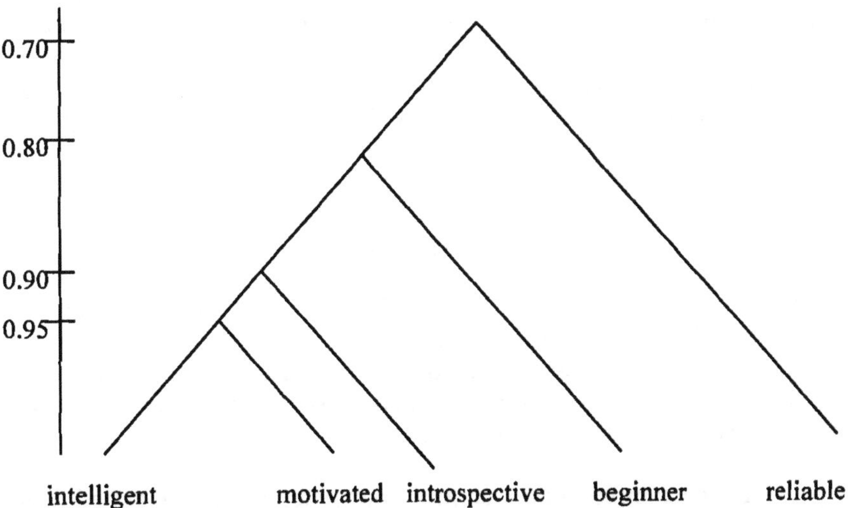

Figure 4.2 Dendrogram for the hierarchical clustering of staff constructs.

4.3 THE KNOWLEDGE LEVEL

4.3.1 Separation of domain and control knowledge, the knowledge level

There are two types of knowledge contained in the knowledge base of an expert system: *domain knowledge* and *control knowledge*. Domain knowledge is specific to the problem at hand. Control knowledge is of more abstract nature and describes overall objectives and strategies of how the system is to pursue a goal. In first generation expert systems, no distinction was made between the two types of knowledge and control knowledge would either be absent or implicit and hidden within the domain specific knowledge. Consider for example the MYCIN rule, written in predicate calculus format, (Clancey, [9]):

$$has(X,infection) \wedge CSFcellcount(X,Y) \wedge Y<=10 \Rightarrow$$
$$has(X,meningitis), CF=-1000$$

There is an important control strategy here which is implicit in the design of the rules. The order of the clauses in the conjunction is not arbitrary. Before checking the CSF cellcount of a patient, it should be the case that the patient has an infection. So, there is an implicit disease hierarchy which determines the rule structure. The general strategy may be described as one of top-down refinement: *before you check the specific, check the general*. This strategy remains unstated in the knowledge base. Making such control knowledge explicit makes it possible to supply more comprehensive explanations of the system decisions, and provide additional guidance into consistently building and maintaining the knowledge base.

In the *knowledge level* model for specification of expert systems, the organization and structure of domain knowledge is made explicit. The knowledge level contains meta-knowledge about how and why control proceeds in a certain way. Many different abstraction levels of meta-knowledge are possible. The separation of domain and control knowledge can make it easier to modify or maintain the knowledge base, can have the effect of building better explanation facilities and can even help the knowledge acquisition process.

As an example of how control knowledge can be separated from domain knowledge, consider the decision tree for the risk classification of an investor contained in Section 7 of Appendix A on Prolog and the domain knowledge rules:

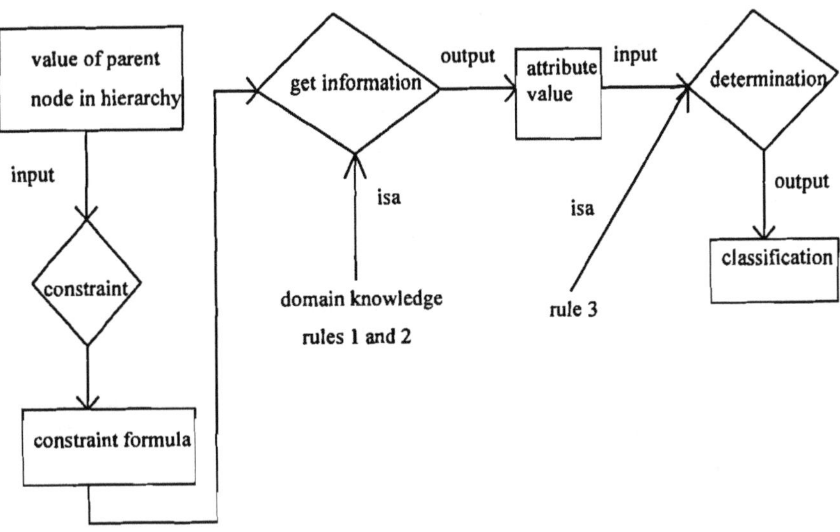

Figure 4.3 Control knowledge for "ask-classify" problems.

IF marital status is married THEN ask number of children.
IF number of children greater than 2 THEN ask income.
IF married and number of children>2 and income<40000 THEN classification=low risk.

The following meta-knowledge is implicit in the domain knowledge given above. The first two rules make sure that the hierarchy in the decision tree is respected in the sense that no child node is asked before its parent. For example, we should not ask the user how many children he has, unless we know that he is married first. The third rule is a classifier which based on values of the parameters chooses the appropriate risk category for the investor.

CommonKADS (Schreiber et al., [40]) is an example of an expert system shell which provides a modeling language to make control knowledge explicit. The meta-knowledge given above can be made explicit in CommonKads using the "conceptual modeler". The diagram in Figure 4.3 shows a graph representation of the meta-knowledge for the investor classification problem.

The control knowledge contained in this figure could be described as follows. The "get-information" type of task finds out which attribute values should be asked from the user next, based on constraints among values of ancestor nodes in the decision tree hierarchy. The "determination" type of task classifies an object based on its attribute values. Observe that both of these descriptions are of a more abstract nature than the domain specific rules and do not contain domain specific items. A second, more abstract level of control knowledge could be constructed describing the strategy of performing the tasks of "get-information" and "determination" in sequence, as shown in Figure 4.4.

When an explanation is requested by the user, it could be given at one of many levels of abstraction. At the lowest level, it will be a retracing of the rule steps. At a higher level, it could explain that taxonomic constraints lead to asking a certain question. Or finally, at the "ask_classify" level, the more general description of "values are solicited based on taxonomic constraints in order to perform classification".

A weakness of the knowledge level modeling approach is that the process of mapping a complex problem domain to an appropriate model is not easy and is not well understood. Also partitioning the knowledge into control and domain knowledge may in itself be difficult. Different components of a complex system may need to be mapped to different models. In that case, all the knowledge components in the system have to communicate results and share inference. Methods are needed for such conversions.

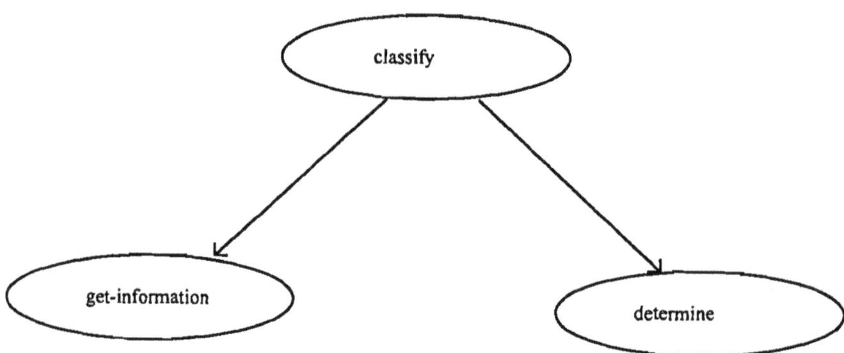

Figure 4.4 A control strategy for "ask-classify".

4.3.2 Neomycin

NEOMYCIN is an example of a system which separates control from domain knowledge. It uses metarules to provide higher level representations for its general problem solving strategy, named *heuristic classification*, and addresses problems in MYCIN's explanation facilities, (Clancey, [9], Hasling and Clancey, [22]).

The following are the major inference steps in heuristic classification: abstracting of data or generalization, matching data to a possible solution through causal relationships, and refining it by subtype classification to the final solution. The hierarchical structure of the domain knowledge is thus made explicit and rules aimed at performing a certain task (for example, get information from the user) are grouped together into rule-sets. For example, FINDOUT is a rule set associated with getting input from the user, EXPLORE-AND-REFINE checks if the current hypothesis is true and if so, it then tries to establish its subtypes (its children in the hierarchy). The metarule for describing EXPLORE-AND-REFINE establishes that:

IF the current hypothesis has a child in the hierarchy and the child has not been verified yet THEN pursue that child.

FINDOUT, the metarule associated with retrieval of information from the user, makes explicit the strategy that, *if the superclass is not verified, then questions about its subclasses should not be asked* (do not ask if patient had brain surgery, if it is not known whether he had surgery).

In Neomycin, a rule in a rule set can fire only if that rule set is active. Metarules provide control across rule sets. Each rule set corresponds to a phase in the system and metarules guide the system from phase to phase. This clustering of rules provides an abstraction which enables NEOMYCIN to provide explanations related to the solution strategy and not just a paraphrasing of the chain of rule firings.

4.4 EXPLANATION IN SECOND GENERATION EXPERT SYSTEMS

One of the advantages of expert systems over alternate techniques for heuristic problem solving, such as neural networks, is that expert systems can provide explanations of how and why the results are reached. The MYCIN experiment showed that providing explanations is crucial to the success and acceptance of an expert system, (Teach and Shortliffe, [47]). A reliable explanation facility can also serve as a verification technique of sorts, since when the system fails to adequately explain its recommendation, the user will be prompted to double-check the result.

In first generation expert systems like MYCIN, explanation was basically a trace of the inference steps. Explanations were provided by para-phrasing the rules involved in the reasoning chain into natural language, using fill in templates, (Scott et al., [41]). This has the inherent weakness that general concepts and strategies needed by the user in order to understand the explanation could not be given. Early systems lacked the level of abstraction necessary to explain a general problem solving methodology. Explanations were often burdened with unnecessary detail clouding the major or important points and would not take into account the user's level of expertise or the context in which an explanation was asked for.

In second generation expert systems, explanation is considered to be a problem solving activity in its own right and meta-knowledge is used to provide an explanation which is understandable, at an appropriate level of abstraction and accepts user's feedback (if user does not understand, the system can further explain by changing the detail level or the perspective of the explanation). Using meta-knowledge, explanations can be given using terms that are more familiar and understandable to the user. Explanations are not static but they can be dynamically generated and the system can adapt its explanations to the characteristics of the individual user.

In addition to CommonKADS, we describe below an explanation generation system developed by Moore et al., [30].

In Moore's system, different levels of explanation from more abstract to more detailed are available to the user by providing the user with the opportunity to ask follow-up questions on the original explanation. Explanations are in natural language and once an explanation is given, the user can highlight the part of the text that he wants further explained.

The explanation system consists of an *application module*, a *text planning module*, a *surface realization module*, a *knowledge base* and the *question generator*. The *application module* builds a user profile based on captured information and available data about the user. The *text planning module* accepts as input a request for information and using the knowledge base constructs a sequence of speech acts to fulfill the request. A *speech-act* is an expression of the form: *(<speech-act> (<predicate><concept1>...<conceptn>))*. In the particular application of advising patients with migraine headaches which is reported in [30], <speech-act> can be one of the following: *inform*, *ask*, *recommend*, and *command*. The *surface realization module* translates these speech acts into natural language text using Functional Unification Grammars, (see [15] or freeware, available from the ftp address: ftp.cs.columbia.edu/pub/fuf). The input to the *question generator* is a highlighted part of the text. The question generator creates a set of follow-up questions in the form of speech-acts that can be asked. Using the user's profile, the history of past questions and answers and the context of discourse, the system prunes the set of questions to a subset appropriate for the individual user. The surface realization module converts these questions into natural language and exhibits them to the user in the form of a menu. The conversion of speech-acts to natural language is done using the *Penman Upper Model* which is a general purpose ontology of linguistic abstractions. For example, the speech-act: (inform is_a ibuprofen analgesic-drug) would be converted to the phrase "ibuprofen is an analgesic-drug". If now the sentence "ibuprofen is an analgesic-drug" is highlighted by the user, the question generator will generate the following set of questions in a menu: Why is it that ibuprofen is an analgesic-drug? What is ibuprofen? What is an analgesic-drug? The user can then pick the question he/she needs further explanation on.

For a detailed description of Moore's explanation system, the reader is referred to [29] and [30].

4.5 THE PROBLEM SOLVING METHODS AND GENERIC TASKS APPROACH

4.5.1 Introduction and analysis

One of the approaches adopted in second generation expert systems, which eases the knowledge acquisition bottleneck, is to classify problem domains according to the types of tasks that are to be performed and build corresponding knowledge level *models* and *problem solving methods* for each type of task. The control knowledge needed to solve problems of a generic type, for example NEOMYCIN's heuristic classification problem, does not depend on particular instances of the problem. Many generally applicable problem solving methods have been identified including *heuristic classification, generate test and debug, propose and revise, cover and differentiate*.

Advantages of building expert systems around problem solving methods include the systematic way in which knowledge acquisition can be contacted. The problem solving method in a sense dictates what type of domain knowledge is required to solve the problem.

Another advantage of using the problem solving methods approach is that explanations can be provided at various abstraction levels and not just by tracing the line of inferencing. Explanations can involve general strategy statements on which phase of the method is being applied.

Among the disadvantages of having the problem solving methods guide the knowledge acquisition process is that the knowledge engineer cannot detect when the knowledge in the system becomes complete, and whether to further pursue or terminate the knowledge acquisition process.

In another but related approach, a methodology of developing expert systems is centered around the analysis of tasks to be performed. Some major classes of tasks have been identified called *generic tasks* (Chandrasekaran, [6,7]). Examples include *diagnosis*, *constraint satisfaction tasks*, *classification* and *constructive tasks*. The way specific tasks are decomposed into simpler subtasks is the same for tasks within a single generic task class. Problem solving methods can be viewed as applying domain knowledge to a task. One weakness of the generic tasks approach is that tasks in a generic class can be performed using alternate problem solving methods and it is not clear which problem solving methods should be associated with a generic task. Another problem area is how to determine which task constitutes a generic task and which does not because it can be decomposed into a set of generic components.

4.5.2 Knowledge acquisition tools for the generic tasks, methods approach

Both automated and semi-automated approaches have been proposed to provide assistance with the knowledge acquisition process. The fully automated approach intends to altogether eliminate the role of the knowledge engineer by using machine learning techniques to automatically derive the knowledge base from raw data. This approach will be examined further in Chapter 6 on Machine Learning and Chapter 7 on Neural Networks. The semi-automated techniques attempt to assist the knowledge engineer with the building of a model for the problem domain.

In this section, we briefly describe some of the semi-automated KA systems proposed along the track of generic tasks, problem solving methods and domain knowledge modeling. Many of them define a certain problem solving methodology for problems (tasks) of a certain type and then ask the user to input the domain knowledge in the format required by the system, while others are broader in scope and allow as input the definition of general problem solving methods and the way to encode the domain knowledge to fit the model. The first type of system is intended to be used directly by a domain expert, while the second requires a degree of sophistication in programming and is intended for use by knowledge engineers. Eventually, it is hoped that this approach will provide a large enough depository of control knowledge methodologies for the various problem types. When faced with a particular application, the domain expert or the knowledge engineer has to choose which problem category the application belongs to and which problem solving method is applicable (for example, is it a heuristic classification or a cover-and-differentiate problem) and use the corresponding tool to help with knowledge elicitation. Of course, there will always be problems difficult to cast as belonging in one of the developed general control methods. Even so, the approach is quite useful, as there will also be a very large number of problems which can be solved by mapping them into one of the problems in the constructed taxonomy of problem solving methods and task domain models.

Examples of knowledge acquisition tools that follow the problem solving methods approach include: MOLE, SALT, SEAR, GT-Toolset and COMMONKADS.

MOLE

MOLE ([26]) solves diagnostic problems by the *cover-and-differentiate* method and is intended for use by the domain expert himself without the need for a knowledge engineer. The cover-and-differentiate method finds possible causes explaining symptoms and then differentiates among the possible causes. The knowledge acquisition module of MOLE asks the user to provide a list of diagnostic problems. For each problem the user is asked to supply possible factors that may have caused the problem and then factors that caused the factors that caused the problem, etc. (the set of problems is thus *covered*). Next the system elicits knowledge to help differentiate among the different possible causes of a problem. In addition to the knowledge acquisition module, the system has its own inference engine, which acts on the knowledge obtained from the expert by the elicitation module. Obviously, for a problem which is a good fit to be solved by MOLE, the set of diagnostic problems or symptoms must be known and possible causes also known and there must be a way to differentiate among causes. For example, the problem "the car doesn't start" can be due to a dead battery or a burned starter. The evidence "head lights do not turn on" helps differentiate the cause in favor of battery being dead. The MOLE has been used among other applications to build expert systems for diagnosing problems in a steel rolling mill, and a coal burning power plant.

SALT

SALT, (Marcus and McDermott, [27]), is a method specific knowledge acquisition tool which performs knowledge elicitation and generates OPS5 type of rules. It implements the *propose-and-revise* method for solving problems that involve *constraint satisfaction tasks*, like planning and configuration tasks. It can be used by the domain experts without the need for a knowledge engineer. Once a problem has been classified as belonging to the constraint satisfaction task type, SALT dictates what knowledge should be acquired. The user is asked to input the following types of knowledge: the parameters and attributes that describe a solution, procedures for assigning values to parameters in the original design, constraints that have to be satisfied by the design parameter values, and procedures on how to modify the parameters when a constraint is violated. In the propose-and-revise method, a plan or design is constructed by proposing a value for a problem parameter and checking if relevant constraints are violated, in which case procedures are applied for adjusting the parameters. Constraints can be identified incrementally as the design continues. For example, we give below a screen that appears in response to the user's selection of "constraint definition" from the SALT menu (from [27]):

1 Constrained value:	CAR-JAMB-RETURN
2 Constraint type:	MAXIMUM
3 Constraint name:	MAXIMUM-CAR-JAMB-RETURN
4 Precondition:	DOOR-OPENING=SIDE
5 Procedure:	CALCULATION
6 Formula:	PANEL_WIDTH*STRINGER_QUANTITY
7 Justification:	PROCEDURE TAKEN FROM MANUAL, p. 12.

This constraint states that the maximum possible value of the *car-jamb-return* parameter is the product of the panel width and the stringer quantity. In the screen above, upper case is what the user typed in as input and lower case are prompts provided by SALT.

SALT has been successfully used for scheduling manufacturing processes and VT is a SALT generated expert system used to design elevator systems at Westinghouse Elevator company.

SEAR

SEAR, ([26]), is a set of tools for developing higher level acquisition tools and its intended user is the knowledge engineer. It allows the knowledge engineer to define his own control strategy methods and how knowledge is to be elicited from the expert.

COMMONKADS

CommonKads, (Schreiber et al., [40]), is an example of an expert system shell which provides a conceptual modeling language to make control knowledge explicit. It allows the user to define his own task knowledge and task control knowledge.

GT-TOOLSET

GT-Toolset is a prime example of the generic task approach and provides models for various generic tasks including classification, hypothesis matching, abduction etc. (Chandrasekaran, [7]). Its task or method specific tools ask the user to formulate the problem to be solved in the systems specific methodology and provide the knowledge of what the system is to achieve. In GT-Toolset, programming is replaced by the specification of problems.

We briefly mention below some other knowledge acquisition tools of interest (not necessarily based on the generic tasks and problem solving methods approach):

TEIRESIAS provides rule and frame editors and hypertext facilities for modeling knowledge, (Davis, [13]).

OPAL is an task-specific acquisition tool used in the cancer chemotherapy domain, (Musen et al., [33]). It helps oncologists build cancer treatment protocols that are used as a knowledge base in the expert system *Oncocin*. It requires users to fill out slots in forms using a predefined vocabulary of representation primitives.

ROSET assists knowledge engineers and domain experts in developing EMYCIN based expert systems for classification problem solving, [3].

ETS, (Boose, [4]), is based on personal construct theory and automates the structured interview and repertory grid analysis. It elicits constructs used by the expert through dialog and builds repertory grids. The grid analysis produces a graphical portrayal of hierarchical relationships between the constructs.

SIS, (Motoda, [31]), is a meta tool for generating specialized knowledge acquisition tools. SIS was used to build IIS-LD, an interviewing system which helps with the logical design of databases.

In Chapter 6, we will introduce the field of Machine Learning, which attempts to reduce or even eliminate the knowledge acquisition bottleneck by deriving knowledge directly from the data.

4.6 VERIFICATION AND VALIDATION

4.6.1 Introduction

Validation of an expert system is concerned with the correctness of the knowledge base and the provision of correct solutions to domain problems. *Verification* is concerned with how appropriate the knowledge representation, explanation facilities and user interfaces are. Verification has as its goal to ensure the consistency and completeness of the knowledge base and to guarantee that the development of the Expert System proceeds in a way that meets the requirements.

Verification and validation of expert systems is much more difficult than in conventional software systems. This is partly due to the fact that an expert system provides solutions in imprecise problem domains by using heuristic, nonalgorithmic techniques. As a result, the solution derived by an expert system is often only an approximation and many alternate solutions may also be acceptable. Another problem is that in contrast to traditional software engineering, there is no external specification against which to test the system. The evolutionary prototyping approach to expert system development makes such specification impossible. This introduces a difficulty in verifying and validating expert systems and makes traditional software engineering verification techniques, such as generate and test, not quite suitable for expert system verification.

As hard as expert system testing may be, the difficulties encountered in knowledge acquisition make the introduction of verification and validation methods even more important for the successful deployment of an expert system. In some application environments, an expert system will not be accepted and can not be put to use unless it can be thoroughly and convincingly tested that performance is according to expectations. Although many open problems still remain, significant progress has been made in the area of expert system verification and validation. Some of the approaches used to validate expert systems include the generation and usage of test cases, evaluation by experts, and comparisons of decisions made by the system to those of experts. In addition to these traditional approaches, various verification techniques have been developed for checking the knowledge base for completeness and consistency, and a number of systems have been built which attempt to automate the verification and validation process ([35, 46]).

4.6.2 Verification

Checking the knowledge base for correctness involves checking it for consistency and completeness. Consistency checking involves testing for *redundant* rules, *conflicting* rules, *subsumed* rules, *circular* rules and presence of *unnecessary rule conditions*. Checking for completeness entails checks for *dead end* rules, *unnecessary conditions* and *dangling* conditions.

4.6.2.1 Inconsistencies

Redundant rules lead to inefficiency, decrease in performance and if uncertainty is present, they can corrupt the uncertainty factor of the derived conclusion. There are two types of redundant rules. A rule is *redundant* if it shares the same conclusion with another rule, and its conjunctive premises are permutations of the conjunctive premises of the other rule. A rule is also *redundant* if it is the transitive closure of a chain of other rules. In other words, given the rule set

$$\{h_1 \Rightarrow h_2,\ h_2 \Rightarrow h_3,\ \ldots,\ h_{n-1} \Rightarrow h_n,\ h_1 \Rightarrow h_n\}$$

the last rule in the set is redundant, since h_n belongs to the transitive closure of h_1.

Conflicting rules have identical conditions but contradicting conclusions. In the case where rules have the same conditions but different conclusions, not necessarily negations of each other, they are to be examined closely. For example, consider the rules

union_member(X)⇒democrat(X) and
union_member(X)⇒republican(X).

If the sets of democrats and republicans are mutually exclusive, then the rules are conflicting and should be adjusted. In the presence of uncertainty, such rules may be valid. For example, given the rule set:

union_member(X)⇒democrat(X), 0.90 and
union_member(X)⇒republican(X), 0.10

it may be that the knowledge engineer intentionally created the two rules. The numeric value at the end of each rule indicates the confidence in the conclusion being true given that the premise is true. A union member is most probably a democrat (with certainty factor of 90 percent), but there is still a small chance that he actually may be a republican (the certainty for that being 10 percent).

Another case of potentially conflicting rules, (and the most difficult case to detect), is the case where the conclusions are contradictory or mutually exclusive, but the conditions are not identical and can potentially happen at the same time. Consider the following example, ([13]):

IF (the red light goes on and reactor temperature is above 1500) ⇒open valve #3
IF (the blue light goes on and reactor temperature is above 1500) ⇒open valve #4
IF (the green light goes on and reactor temperature is above 1500) ⇒close valve #4.

What if the blue and green light are both on? Are the events "blue light", "green light" mutually exclusive or is it a rule design fault?

A rule R_1 is *subsumed* by a rule R_2 iff they share the same conclusion and the set of conditions of rule R_2 is a subset of the set of conditions of rule R_1. A subsumed rule can be deleted, since when it succeeds the more general rule also succeeds. For example, consider the rules

fisherman(X) $\wedge$ *male(X)* $\Rightarrow$ *has_boat(X)*
and
fisherman(X) $\Rightarrow$ *has_boat(X).*

The first rule is subsumed by the second and can be deleted.

Again, in the presence of uncertainty subsumed rules cannot be deleted as they may be intentional. For example, consider the rule:

If the textile is hydrophobic THEN it is nylon, with confidence factor 0.3

and the rule:

If the textile is hydrophobic and wrinkle resistant THEN it is nylon, with confidence factor 0.7.

The second rule indicates that in the presence of more evidence our belief to the textile being nylon increases.

Circular rules are of the form:
If a_1 then a_2, If a_2 then a_3, ..., If a_n then a_1.

The existence of circular chains of rules may cause an infinite

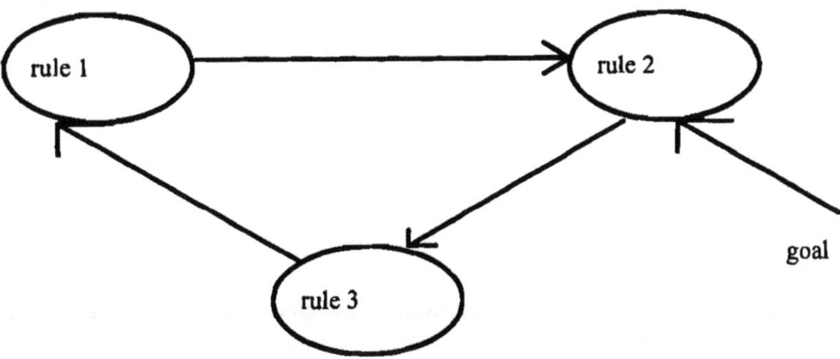

Figure 4.5 A rule dependency graph.

loop during the inferencing process. A *rule dependency graph* can be used as a method for detecting cycles in the rule set. For example, consider the rules:

(R_1) $a(X) \Rightarrow b(Y)$
(R_2) $b(Y) \wedge c(Z) \Rightarrow g$
(R_3) $g \Rightarrow a(X)$

This rule set can be represented by the dependency graph of Figure 4.5.

The nodes of the dependency graph correspond to rules and one of them is designated as the goal node. A directed arc connects the node for a rule R to the node for a rule S iff the conclusion of S is subsumed by the condition of R. Circular chains of rules are detected by starting at the goal node and traversing the graph to see if there is a cycle.

4.6.2.2 Completeness

Unnecessary conditions.
Consider the rules: (R_1) $a \wedge b \Rightarrow c$ and (R_2) $a \wedge \neg b \Rightarrow c$. Then the truth value of b does not depend on the validity of c, and the set of the two rules given above can be substituted by the single rule: $a \Rightarrow c$. Again, in the presence of uncertainty, a rule set like (R_1) and (R_2) may both be needed to express the different certainties with which we can conclude c, depending on the truth value of b.

Dead-end rules
A rule, *If condition then conclusion*, is a dead end rule if its conclusion is not a goal predicate (a candidate query) and is not used as a condition of any other rule. Then this rule can never trigger another rule and it is either unnecessary or some other rule using its conclusion as hypothesis is missing.

Dangling conditions
A rule whose condition will never be reached as a conclusion by another rule and does not match any fact is called an *unreachable* rule. Its condition is called a dangling condition and can never be verified, so the rule conclusion can never be reached. The presence of a dangling condition may indicate that some other rules or facts have been erroneously omitted.

The presence of inconsistencies and incompleteness in a knowledge base does not necessarily reveal errors, since the knowledge engineer may have semantically intended the structure. As a result, the function of a verification and validation algorithm should be to exhibit potential errors for consideration by the domain expert or the knowledge engineer, and not to undertake corrections of suspected errors itself.

Various algorithms for checking the knowledge base of an expert system for consistency and completeness can be found in Nguyen et al., ([35]). TEIRESIAS, (Davis, [13]), was one of the first systems to help the knowledge engineer in knowledge verification. CHECK, (Nguyen, [35]), is another validation tool used to check knowledge bases created by the LES (Lockheed Expert System) shell. CHECK checks for the completeness and consistency situations described above for both forward and backward chaining inference and also for knowledge bases with certainty factors. Other verification tools include DARC, ([36]), and Ginsberg's KB-REDUCER, ([21]). Several commercial expert system shells include knowledge base verification components. For example, EXSYS checks for any conflicting or identical rules in the knowledge base whenever a new rule is added.

4.6.2.3 Verification using Petri nets

The predicate transition net (PTN) introduced by Genrich, ([20]), is a variation of the basic Petri net model and can be used to model knowledge bases. In Chapter 4, Section 7, we presented a methodology of how to transform a set of rules to an equivalent Petri net and vice versa.

In the Petri net formalism, a knowledge base is logically inconsistent (the empty clause can be derived from it) iff a place eventually holds a black and a white token simultaneously.

For example, consider the following knowledge base:

fly(X) $\Rightarrow$ bird(X).
cat(X) $\wedge$ bird(X) $\Rightarrow$ eat(X,Y).
$\neg$eat(tom, tweety).
cat(tom).
fly(tweety).

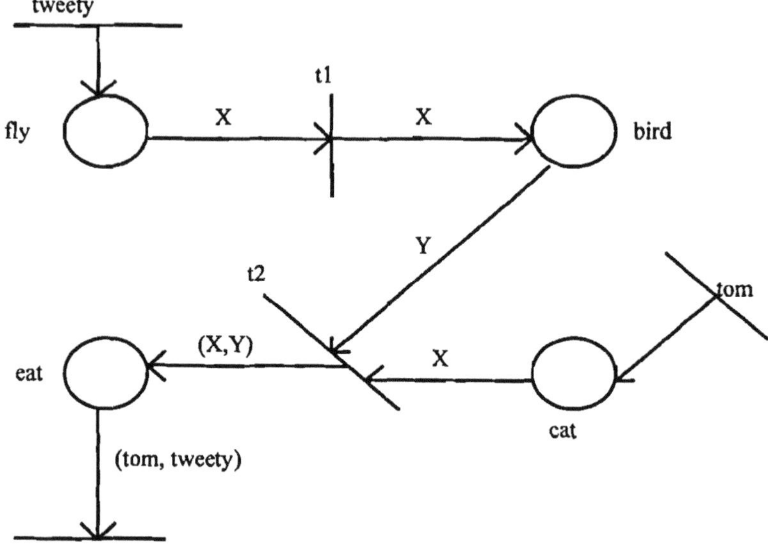

Figure 4.6 A Petri net representation of the cat, bird knowledge base.

The equivalent Petri net is shown in Figure 4.6.

Each fact corresponds to a token. Corresponding to the fact *¬eat(tom, tweety)*, the black token *(black, tom, tweety)* is deposited in the place labeled by the predicate *eat*. Similarly, each of the remaining facts corresponds to a token. The token *(white, tom*) is deposited in the place *cat* and the token *(white, tweety)* in the place *fly*.

Transition 1 is activated and deposits the token *(white, tweety)* in place *bird*. Transition 2 is enabled and fires to deposit the token *(white, tom, tweety)* in the place *eat*. As a result, the place node corresponding to the predicate eat contains the tokens *(black, tom, tweety)* and *(white, tom, tweety)*. Therefore the knowledge base is inconsistent.

The Petri net model of a knowledge base can also be searched to verify a knowledge base for incompleteness and inconsistency in terms of redundant, subsumed, circular, conflicting and dangling condition rules. Each one of these types of rules corresponds to a pattern subnet of the original net. Their presence in a knowledge base can be identified by searching the Petri net representation for certain patterns of subnets. Figure 4.7 shows the Petri net subnet patterns corresponding to the two

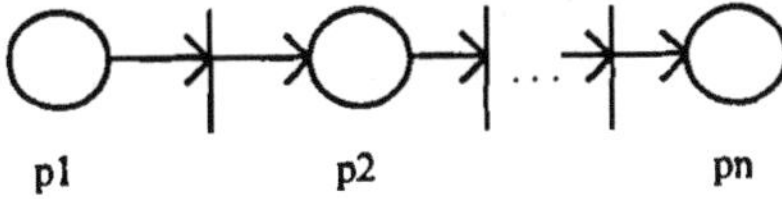

(a) Redundant rule p1 => p2

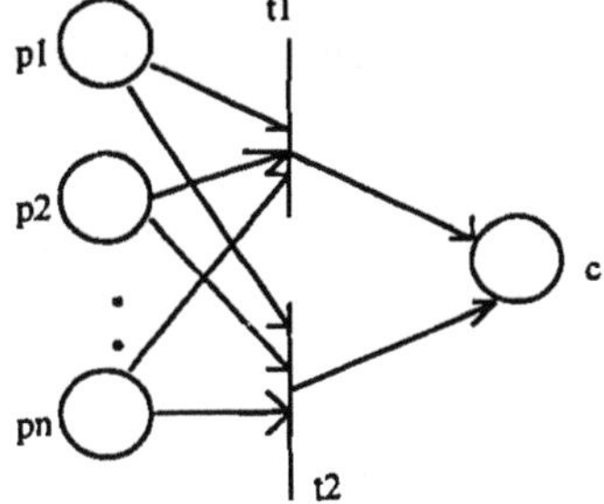

(b) Redundant rule in transition t2

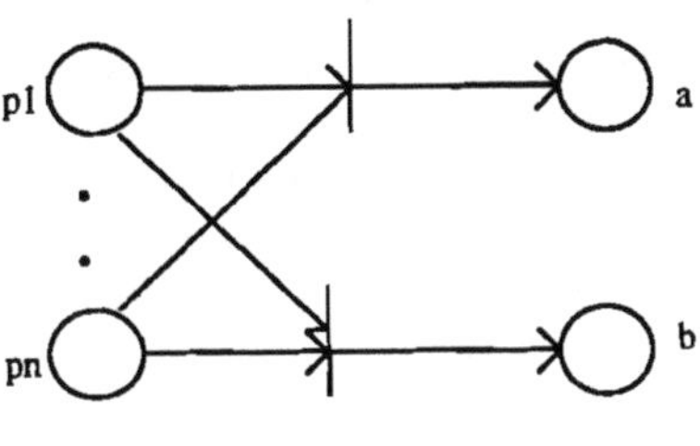

(c) Conflicting rule pattern

Figure 4.7 Inconsistency pattern subnets.

types of redundant rules and conflicting rules.

In Zhang and Nguyen, ([50]), each inconsistent pattern is expressed as an inconsistency string grammar. For example, if a place node is denoted by the symbol a, a transition by the symbol t, an arc incoming to a transition by b and an outgoing arc by d, then the grammar defining a redundant or subsumed rule contains the rules:

{S→A, A→AA, A→BC, B→bD, D→aE, E→b, F→d, C→GH, G→bG, G→Ia, I→tF, H→Ft}, where S, A, B, C, D, E, F, G, H, I are nonterminal symbols.

Incomplete rules are detected by checking each place node (predicate) in the Petri net to see if it has no incoming or outgoing arcs. Rules involving such predicates are potentially erroneous rules.

Inconsistencies are found by searching the net for *closed connections*. A closed connection is a subnet whose underlying graph is a cycle. Not every closed connection subnet is an inconsistent pattern, but the set of closed subnets provides the set of possible inconsistent patterns.

Each closed connection is translated into a string pattern and parsed using the inconsistency pattern grammars to check if it is truly an inconsistent pattern.

The set of all possibly incomplete or inconsistent rules can then be made available to the expert for evaluation.

4.6.3 Validation

In expert systems, the knowledge base is often based on heuristics and nonalgorithmic knowledge, so the question of validation is not just about the output being correct or not. Validation of an expert system involves trying to substantiate that the system performs with an acceptable level of accuracy comparable to human performance. After the knowledge base has been verified, validation makes sure that the expert system correctly models the decision process. In the same way as verification, validation is performed throughout the development cycle.

4.6.3.1 Validation of MYCIN

During the validation of MYCIN, (Buchanan, [5]), faculty at the Stanford division of Infectious Diseases were given test cases and asked to evaluate MYCIN's suggested treatment, ability to identify the microorganisms causing the disease, as well as overall performance. MYCIN received an approval rating of 75%. To eliminate the possibility of bias, a second experiment was performed, in which experts from various universities across the country were chosen and again performance was evaluated. It remained at 75%. A third study was done, where a variation of the *Turing test* was used, (Yu et al., [49]). Diagnosis and treatment plans for the same set of patients were produced independently by MYCIN and a host of experts. The results were presented to another group of experts without specifying to them which results were derived by MYCIN and which by humans. The second group of experts were then asked to rank each diagnosis and treatment, as either an acceptable alternative to what they would prescribe or unacceptable. MYCIN outperformed the human experts of the first group suggesting equivalent or acceptable diagnosis to 65% of the cases.

4.6.3.2 Testing

The MYCIN experiment addressed a number of validation issues. Experts or developers may have certain biases, which they may even be unaware of, and which may influence the validation process (an expert may resent the attempt to have computers introduced in his domain or a developer may want the system to be a success, so he chooses favorable test cases). To avoid bias, the experts and developers involved in developing the system should not be the only ones validating the system.

The system performance can be checked by presenting test cases to the system. The cases to be used for testing should not have been used during the knowledge acquisition and development phases. Two types of testing are possible. In *black box testing*, the results from the expert system can be shown to one or more experts for critique. In *white box testing*, the results, together with explanations on why the particular result was reached, are provided to the experts. The experts can rate the results, for example from 0 (total disagreement) to 5 (total agreement), and based on the ranking, a qualitative determination of the system's accuracy can be made. Alternatively the test cases can be presented to the expert system and the human experts at the same time and results compared.

To avoid biases, in addition to the above, a variation of the *Turing test* can be applied, as was done in MYCIN. In this test, the results obtained by a human expert and by the expert system are presented in the same format to expert evaluators, without informing them of which was derived by the machine and which by the human. If evaluators cannot distinguish the answers of the expert system from those of the human expert, then the expert system passed the test.

Field testing is another common methodology used in the validation process. The system is placed on various sites and made available to its intended users. This way, the total package, including the user interface, can be evaluated in a real working environment.

The user interface design of an expert system is quite important as it affects the acceptability of the expert system by its intended users. The needs and the profile of the user must be taken into account during design. For example, whether the users are novices or experts in the application domain determines at which level and how extensive the explanations need to be. Current expert system interfaces employ menus, graphics, icons, hypertext and natural language interfaces among the various technologies currently available.

When test cases are available for which both the system and the expert's solutions are available, validity can be tested statistically. For example, in a paired t-test, the system is valid iff 0 belongs to the confidence interval given by $m \pm t_{n-1,\alpha/2}S_m/\sqrt{n}$, where m is the mean of the differences d_i=(system result or performance)-(expert's result or performance), for each test case i=1,...,n. S_m is the standard deviation of m and $t_{n-1,\alpha/2}$ the t-distribution with n degrees of freedom.

It seems to be difficult to combine performance measures in different areas in one aggregate measure. Various methods of multicriteria analysis have been applied for a set or a single aggregate measure. The reader is referred to Cohon, ([11]), for more information on performance measuring.

An inherent weakness of the test-case approach is that as the number of rules grows, the number of the test cases which are needed to exhaustively test the system grows exponentially. Systems such as COVADIS, ([38]), and MELODIA, ([8]), check for incompleteness of knowledge by identifying cases which are not considered by the knowledge base and for external validity problems by identifying cases which can lead to problematic situations. The identified cases are then presented to the human expert who decides if they are important suggesting a problem in the knowledge base or not.

SUMMARY

Knowledge acquisition is a major bottleneck in expert system development. In second generation expert systems, knowledge acquisition is approached as a modeling activity and not just as a knowledge transfer from the expert to the computer via the knowledge engineer. This chapter introduced various knowledge elicitation, verification and validation techniques for expert systems. The *knowledge level* approach to expert system development was introduced and how this approach is employed in second generation expert systems was examined. The generic tasks and problem solving methods approach and its effects on second generation systems was also covered.

EXERCISES

1. Consult with an instructor in one of your classes to elicit knowledge on what characteristics are important for a student to be successful in the class. Ask your instructor to pick a representative set of students (not named but labeled 1 to k) and a set of bipolar constructs which influence a student's success.
   a) Construct a repertory grid using the students as elements of discourse and the ratings provided by the instructor on how each student fits the constructs.
   b) Use the hierarchical clustering technique to cluster the elements in the domain of discourse, as well as the set of constructs. Create a focused grid, show it to the instructor and consult to see if it prompts any change in the knowledge from his part on the constructs' importance, relevance, etc.
   c) Use NICOD's theory of confirmation to derive entailments between different constructs.

2. Consider the set of rules:

   dept_in(electricalengineering, engineering).
   dept_in(civilengineering, engineering).
   ¬status(electrical_engineering, accredited).
   status(engineering, accredited).
   dept_in(X,Y), status(Y,Z)⇒status(X,Z).

   a) Construct a corresponding Petri net and
   b) Use the Petri Net to show that the rule set is contradictory.

REFERENCES

[1] Anderson, J.R., "Skill Acquisition: Compilation of weak method problem solutions", *Psychological Review*, 94, 1987, pp. 192-210.

[2] Batty, D., Kamel, M., "Automating Knowledge Acquisition", *IEEE Trans. on Knowledge and Data Engin.*, vol. 7, no. 1, 1995, pp. 53-67.

[3] Bennett, J.S., "ROGET: A knowledge based system for acquiring the conceptual structure of a diagnostic Expert System", *Journal of Automated Reasoning*, 1, 1985, pp. 49-74.

[4] Boose, J.H., "A Knowledge acquisition program for Expert Systems based on personal construct psychology", *Intern. J. of Man-Machine Studies*, 23, 1985, pp. 495-525.

[5] Buchanan, B.G., Shortliffe, E.H. "Uncertainty and Evidential support", in *Rule Based Expert Systems: The MYCIN Experiments of the Stanford Heuristic Programming Project*, eds. Buchanan and Shortliffe, Addison Wesley, 1984, pp. 209-232.

[6] Chandrasekaran, B., "Towards a taxonomy of Problem-Solving types", *AI Magazine*, 4, 1, pp. 9-17.

[7] Chandrasekaran, B., "Towards a functional architecture for intelligence based on generic information processing tasks", *Proceedings 10th IJCAI*, Milano, Italy, 1987.

[8] Charles, E., Dubois, O., "MELODIA: Logical Methods for Checking Knowledge Bases", *Validation, Verification and Test of Knowledge Based Systems*, Ayel, M., Laurent, J.P., eds., Chichester, John Wiley, 1991.

[9] Clancey, W., "The epistemology of a rule based expert system: A framework for explanation", *Artificial Intelligence*, 20, 3, 1983, pp. 215-251.

[10] Clark, T., "Language use and language users", In *Handbook of Social Psychology*, Addison Wesley, 1985, pp. 179-232.

[11] Cohon, J.L., *Multiobjective Programming and Planning*, Academic Press, New York, 1978.

[12] Cooke, N.J., "Varieties of Knowledge Elicitation Techniques", *Int J. Human-Computer Studies*, 41, 1994, pp. 801-849.

[13] Davis, R., "Interactive Transfer of Expertise: Acquisition of New Inference Rules", *Artificial Intelligence*, vol. 12,#2, 1979, pp. 121-157.

[14] Dubes, R., *Algorithms for Clustering Data*, Prentice Hall, 1987.

[15] Elhadad, M., Robin, J., "Controlling Content Realization with Functional Unification Grammars", *Proceedings 6th Intern.*

Workshop Natural Language Generation, Springer-Verlag, 1992.

[16] Ericsson, K.A., Simon, H.A., *Prorocol Analysis:Verbal Reports as Data*, Cambridge, MA, MIT Press, 1984.

[17] Feigenbaum, E.A., "Themes and case studies of knowledge engineering", In: *Expert Systems in the Micro-Electronic Age*, Ed. D. Michie, Edinburg Univ. Press, 1979, pp. 3-25.

[18] Ford, K., Petry, F., Adams-Webber, J. Chang. P., "An approach to knowledge acquisition based on the structure of personal construct systems", *IEEE Trans. on Knowledge and Data Engin.*, vol. 3, 1, 1991, pp. 78-88.

[19] Gaines, B.R., Shaw, M.L.G., "Introduction of inference rules for expert systems", *Fuzzy Sets and Systems*, vol. 18, Elsevier, 1986, pp. 315-328.

[20] Genrich, H.J., "Predicate/Transition Nets", in *Lecture Notes in Computer Science*, No. 254, Springer-Verlag, 1987, pp. 207-247.

[21] Ginsberg, A., "Knowledge Base Reduction: A New Approach to Checking Knowledge Bases for Inconsistency and Redundancy", *Proc. Seventh Nat. Conf. Artif. Intel.*, 1988.

[22] Hasling, D.W., Clancey, W.J., Rennels, G., "Strategic explanations for a diagnostic consultation system", *International Journal of Man-Machine Studies*, 20, 1, 1984.

[23] Hoffman, R., "The problem of extracting the knowledge of experts from the perspective of experimental psychology", *AI Magazine*, 8,2, 1987, pp. 53-67.

[24] Kelly, G., A., *The Psychology of Personal Constructs*, NY, Norton, 1955.

[25] LaFrance, M., "The Knowledge Acquisition Grid: a method for training knowledge engineers", In: *Knowledge Acquisition for Knowledge Based Systems*, (Eds.), Gaines, Boose, Vol. 1, 1988, pp. 81-91.

[26] Marcus, S., *Automating Knowledge Acquisition for Expert*

Systems, Kluwer Academic, 1988.

[27] Marcus, S., McDermott, J., "SALT: A knowledge acquisition tool for propose and revise systems", *Artificial Intelligence*, 39, 1989, pp. 1-37.

[28] McCarthy, J., Hayes, P.J., "Some philosophical problems from the standpoint of AI", In: *Machine Intelligence*, vol. 4, Edinburg Univ. Press, 1969, pp. 463-502.

[29] Moore, J.D., *Participating in Explanatory Dialogues: Interpreting and Responding to Questions in Context*, MIT Press, Cambridge, Mass., 1995.

[30] Moore, J.D., Mittal, V., "Dynamically Generated Follow-up Questions", *Computer*, 29, 7, July 1996, pp. 75-86.

[31] Motoda, H., "The Current status of Expert System Development", *IEEE Expert*, 5, 4, Aug. 1990, pp. 3-11.

[32] Musen, M.A., Fagan, L.M., Combs, D.M., Shortliffe, E.H., "Use of a domain model to drive an interactive knowledge editing tool", *Intern. J. of Man-Machine Studies*, vol. 26, #1, 1987, pp. 105-121.

[33] Myer, M. Booker, J., *Eliciting and analyzing Expert judgement: A practical guide*, Academic Press, 1991(cluster analysis techniques in KA).

[34] Newell, A., "The Knowledge Level", *Artificial Intelligence*, 18,1, 1982, pp. 87-127.

[35] Nguyen, T.A., W.A. Perkins, Laffey, T.J., Pecora, D., "Knowledge Base Verification", *AI Magazine*, Summer 1987, pp. 69-75.

[36] Preece, A.D., Shinghal, R., "DARC: A Procedure for Verifying Rule-based Systems", *Proc. World Congress on Expert Systems*, 1991.

[37] Regoczei, S., Plantinga, E.P.D, "Creating the domain of discourse: Ontology and inventory", *Intern. J. of Man-Machine Studies*, 27, 1987, pp. 235-270.

[38] Rousset, M.C., "On the Consistency of Knowledge Bases: The COVADIS System", *Proc. European Confer. Artif. Intell.*, ECAI-88, 1988, pp. 79-84.

[39] Rugg, G., Shadbolt, N., "On the limitations of repertory grids in knowledge acquisition", *Proceedings 1991 Banff Knowledge Acquisition Workshop.*

[40] Schreiber, G., Wielinga, B., DeHoog, R., Akkermans, H., Van de Velde, W., "CommonKADS: A Comprehensive Methodology for KBS Development", *IEEE Expert*, vol. 9, n. 6, Dec. 1994, pp. 28-37.

[41] Scott, A.C., Clayton, J.E., Gibson, E.L., *A practical guide to Knowledge Acquisition*, Reading, MA, Addison Wesley, 1991.

[42] Scott, A., Clancey, W., Davis, R., Shortliffe, H., "Methods for generating explanations", In Buchanan, B. and Shortliffe, E.H. (eds), *Rule Based Expert Systems*, Addison Wesley, 1984.

[43] Shaw, M.G., *On becoming a personal scientist*, Academic Press, London, 1980.

[44] Shaw, M.G., *Recent Advances in Personal Construct Technology*, Academic Press, 1981.

[45] Shaw, M.G., "Planet: Some experience in creating an integrated system for repertory grid applications on a microcomputer", *Intern. J. Man-Machine studies*, vol. 17, 1982, pp. 345-360.

[46] Sowa, M., Scott, A.C., Shortliffe, E., "Completeness and Consistency in a rule based system", in *Rule Based Expert Systems*, Ed. B.G. Buchanan and E.H. Shortliffe, Addison Wesley, 1985.

[47] Teach, R., Shortliffe, E., "An Analysis of physician's attitudes", In: *Rule Based Expert Systems: The MYCIN Experiments of the Stanford Heuristic Programming Project*, Reading, MA, Addison Wesley, 1984.

[48] Winograd, T., Flores, F., *Understanding Computers and Cognition: A new foundation for design*, Norwood, NJ, Ablex, 1986.

[49] Yu, V.L., Fagan, L.M., Bennett, S.W., Clancey, W.J., Scott, A.C., Hanigan, J.F., Blum, R.L., Buchanan, B.G., Cohe S.N., "An evaluation of MYCIN's advice", In: *Rule Based Expert Systems: The MYCIN Experiment of the Stanford Heuristic Programming project*, eds. Buchanan and Shortliffe, Addison Wesley, 1984, pp. 589-596.

[50] Zhang, D., Nguyen, D., "A technique for knowledge base verification", Proceedings *IEEE Intern. Workshop on Tools for A.I.*, IEEE Computer Society Press, 1989, pp. 399-406.

CHAPTER 5

UNCERTAINTY

5.1 UNCERTAINTY IN EXPERT SYSTEMS

In complex real world problem domains, the expert's knowledge, perception of the world and reasoning processes can not always be modeled using boolean logic and boolean inference techniques. Uncertainty is present in many domains, and it can manifest itself as a result of incomplete information, conflicting information, imprecision or vagueness. Errors in the data can also manifest themselves due to erroneous data entry or erroneous sensor readings. Many domains are inherently imprecise and vague. In such domains, conflicting data can also be present due to the presence of multiple data collecting agents or because multiple experts can have different or even conflicting opinions as to the appropriate solution. Under these circumstances, the performance of an expert system to a large extent depends on the way uncertainty is managed by the system.

The boolean, monotonic approach to reasoning does not effectively solve problems in domains involving uncertainty. As a result, a number of alternate theories have been developed to manage uncertainty in expert systems. The majority of such theories are quantitative in nature. In such approaches, a scheme is proposed on how to introduce measures, which numerically quantify uncertainty, and how to propagate and combine these numeric measures of uncertainty during reasoning. In addition, qualitative, nonnumeric theories on the treatment of uncertainty have been proposed. The reader is referred to the theory of endorsements, (Cohen, [10]) and variants of nonmonotonic logic (see Chapter 2).

Uncertainty may be present with respect to the degree of validity of either rule conditions or facts. In addition, rules themselves can be uncertain. For example, consider the following rule, which describes an investment expert's heuristic problem solving knowledge:

IF (investor is middle-aged and income is low) THEN (investor fits the low-risk investor profile).

The expert may be uncertain of the rule itself (usually, investors with these characteristics are low-risk investors, but not always). A reliability factor of 90% may be assigned to the rule which can be interpreted to mean that if the investor satisfies the rule conditions, then he is classified as low-risk with certainty of 90%. Uncertainty could also be present about whether or not a rule condition is met. It may be only 80% certain that the investor's income is low. In that case, what is the certainty of fitting the low risk investment profile? It should be less than 90%, but how are the uncertainties to be combined? Another factor contributing to uncertainty is that the terms "middle-aged" or "income is low" are inherently fuzzy and vague. How are such imprecise linguistic terms to be handled? If a specific interval is used to define the term "middle-aged", for example [45, 55], then what about someone who is 44 years, 11 months and 29 days old? The rule will not be applicable to this investor. The predicate *"age"* has a continuous interval as its domain, and discretisizing this interval into subintervals corresponding to the concepts of young, middle aged, etc., introduces the problem of handling the boundary points.

This chapter concentrates mainly on the quantitative methods which can be employed to deal with uncertainty in Expert Systems, including the Bayesian, Certainty Factors, Dempster-Shafer, Fuzzy Sets and Belief Networks approaches.

5.2 THE BAYESIAN APPROACH

The conditional probability, P(H/E), of events E and H, can be viewed as a quantification of the causal/effect relationship between E and H, with event E viewed as evidence supporting hypothesis H. A probability measure can be interpreted in an objective or subjective way. The objective interpretation of a probability measure is based on frequency counts (number of occurrences of the event divided by the number of trials). When data is not available or the number of available data is not adequate to calculate a frequency count, and the event cannot be repeated a large number of times to generate such data, a different interpretation can be given to probabilities. In such a case, the probability assigned to an event is of subjective nature, and it measures the degree of belief that the person, who assigned the probability, has to the truth of the event. Bayes's theorem provides a formula which can be used to perform probabilistic reasoning in expert systems in the presence of uncertainty.

5.2.1 Bayes' Theorem

The conditional probability of a hypothesis h, given the presence of evidence e, is given by:

$$P(hypothesis/evidence)=P(h/e)=P(h \wedge e) / P(e)=P(h)P(e/h) / P(e)$$

Similarly, for two pieces of evidence:

$$P(h/e_1 \wedge e_2)=P(h)P(e_1 \wedge e_2 /h) / P(e_1 \wedge e_2)$$

and in general:

$$P(h / e_1 \wedge e_2 \wedge ... \wedge e_n)=P(h)P(e_1 \wedge .. \wedge e_n / h) / P(e_1 \wedge ... \wedge e_n)$$

This is known as *Bayes's rule*.

The *multiplicative law* in probability theory states that:

$$P(e_1 \wedge e_2 \wedge .. \wedge e_n)= P(e_1 / e_2 \wedge .. \wedge e_n)P(e_2 / e_3 \wedge .. \wedge e_n)...P(e_{n-1} /e_n)P(e_n).$$

If $h_1, h_2,...,h_k$ are mutually exclusive and exhaustive hypotheses ($P(h_i \wedge h_j)=0$ and $\sum P(h_i)=1$), then a more general form of the Bayesian formula can be derived as follows:

Theorem:
If $P(e_i / e_j)=P(e_i)$ (events e_i are independent) and $h_1, h_2,...,h_k$ are mutually exclusive and exhaustive, then

$$P(h_i /e_1 \wedge ... \wedge e_n)= [P(h_i)P(e_1 /h_i)...P(e_n /h_i)]/[\sum_j P(h_j)P(e_1 /h_j)...P(e_n /h_j)]$$

Proof
We will prove the formula for k=2, h_1=h and h_2=not(h). Namely, we prove that:
$P(h/e_1 \wedge .. \wedge e_n)=[P(h) P(e_1 /h) ... P(e_n /h)] / [P(h) P(e_1 /h)...P(e_n /h)+ P(e_1 /not(h)) ... P(e_n / not(h))(1-P(h))]$.
We have that $P(e_1 \wedge ... \wedge e_n /h)= P(e_1 \wedge ... \wedge e_n \wedge h) / P(h)$. By the multiplicity law, this is equal to $P(e_1 / e_2 \wedge ... \wedge e_n \wedge h)P(e_2 / e_3 \wedge ... \wedge e_n \wedge h)...P(e_n / h)P(h) / P(h)= P(e_1 / e_2 \wedge ... \wedge e_n \wedge h) P(e_2 / e_3 \wedge ... \wedge e_n \wedge h)... P(e_{n-1} / e_n \wedge h)P(e_n /h)$. Since the e_i's are independent, then for each i and for each j≠i, the knowledge that event e_j is true does not affect the truth

of event e_i, and so:
$P(e_i /h)=P(e_i / e_{i+1} \wedge h)=...= P(e_i / e_{i+1} \wedge e_{i+2} \wedge..\wedge e_n \wedge h)$.

As a result the previous equation becomes:
$P(e_1 \wedge e_2 \wedge ... \wedge e_n /h)= P(e_1 /h)P(e_2 /h)...P(e_n /h)$. Also $P(e_1 \wedge..\wedge e_n)=P(e_1 \wedge ...\wedge e_n \wedge h)+ P(e_1 \wedge..\wedge e_n \wedge not(h))= P(e_1 \wedge..\wedge e_n / h)P(h)+ P(e_1 \wedge..\wedge e_n / not(h))P(not(h))= P(e_1 \wedge..\wedge e_n / h)P(h)+ P(e_1 \wedge..\wedge e_n/not(h)) [1-P(h)]$. Therefore the formula is true for k=2. Induction can now lead us to rewriting *Bayes's rule* in the form given in the theorem. QED

Bayes' rule expresses the *a priori* probabilities in terms of *posterior* probabilities, which in certain cases are easier to obtain or estimate. For example, consider the domain of car insurance underwriting and the rule

IF (applicant is good-driver and applicant is a senior) THEN (issue a low rate policy)

By checking the company's records, we can approximate the probability that an applicant is a good driver given that a policy was issued. It would be more difficult to estimate the probability that a policy was issued, given that the applicant is a good driver. For the last case, frequency counts can not be used, since there are no available records.

5.2.2 An example of applying the Bayesian approach

Consider the evidence e_1=single, e_2=high_income, e_3=young, supporting the hypotheses h_1=high_risk_investor or h_2=low_risk_investor, which are mutually exclusive and exhaustive, i.e. $P(h_1 \wedge h_2)=0$ and $P(h_1)=1-P(h_2)$.

Assume that the domain expert estimates the posterior probabilities as:

$P(h_1)=0.3$, $P(h_2)=0.7$, $P(e_1 /h_1) = 0.6$, $P(e_1 /h_2) = 0.3$, $P(e_2 /h_1) = 0.2$, $P(e_2 /h_2) = 0.8$, $P(e_3 / h_1) = 0.5$, $P(e_3 / h_2) = 0.2$.

This could be done by the financial expert by looking at his/her files, and counting the frequency with which investors maintain high_risk portfolios. As an alternative, without using frequency counts the expert can subjectively estimate what is the probability that an investor will choose high risk investments.

The a priori probabilities can now be calculated as:

$P(h_1/e_1)=[P(h_1)P(e_1/h_1)]/[P(h_1)P(e_1/h_1)+P(h_2)P(e_1/h_2)]=(0.3\times0.6)/(0.3\times0.6+0.7\times0.3)= 0.4615$

Similarly,

$P(h_1 /e_2) = 0.097$, $P(h_1 /e_3)=0.517$, $P(h_2 /e_1) = 0.538$, $P(h_2 /e_2) = 0.903$, $P(h_2 /e_3) = 0.483$

Also,

$P(h_1/e_1\wedge e_3)=0.681$, $P(h_2 /e_1\wedge e_3)=0.318$ and
$P(h_1 /e_1\wedge e_2\wedge e_3)= [P(h_1)P(e_1/h_1)P(e_2/h_1) P(e_3/h_1)] / [P(h_1)P(e_1/h_1)P(e_2/h_1)P(e_3/h_1) +P(h_2)P(e_1/h_2)P(e_2 / h_2)P(e_3 / h_2)] =0.349$

and, in the same way: $P(h_2 /e_1\wedge e_2\wedge e_3)= 0.651$

These results are shown in Table 5.1.

Table 5.1 Hypothesis/evidence Bayesian Table.

		After evidence				
	initial probability	e_1	e_2	e_3	$e_1 \wedge e_3$	$e_1\wedge e_2\wedge e_3$
h1=*high risk*	0.30	0.461	0.097	0.517	0.681	0.349
h2=low risk	0.70	0.538	0.90	0.488	0.318	0.651

Based on these results, h_2 is the most probable answer, given that all three pieces of evidence are present. If e_1 and e_3 are present then the most probable answer is h_1, etc.

5.2.3 Weaknesses

The use of the Bayesian approach has the following weaknesses:

1. First it requires knowledge of a very large number of probabilities, namely $P(h_i)$ and $P(e_j/h_i)$ for each evidence e_j and hypothesis h_i, i=1,...,k and j=1,...,n. These (k+1)n probability values may not always be easy to estimate.

2. The formula is only valid under the simplifying assumption that the antecedents e_i are independent. This is not always the case.

3. The probabilistic approach assumes that the presence of evidence also affects the negation of the conclusion. For example, if P(c/a)=0.8, then use of probability theory implies that P(not(c)/a)= 0.2. This is not necessarily true in most domains. For example, in the medical domain the presence of a symptom should not contribute in the evidence supporting both the presence and absence of a disease.

4. If the prior and posterior probabilities are based on frequency counts and statistics, then the samples must be of large enough size to derive accurate probabilities. If the probabilities are not based on frequencies, but are estimated by human domain experts, they may be inconsistent. For example, they may not sum up to one, even if cases are exhaustive.

5. A Bayesian based system cannot easily provide explanation facilities.

Belief networks have been introduced as a means for eliminating evidence is independent. In belief networks, evidence for a hypothesis is not simultaneously thought of as evidence for the negation of the hypothesis. Bayesian belief networks are introduced in Section 5.6.

5.3 CERTAINTY FACTORS

5.3.1 Definitions

The confidence (certainty) factors approach was originally introduced in the MYCIN expert system, (Buchanan and Shortliffe, [8]). MYCIN was one of the earliest expert systems developed and was used to diagnose infectious blood diseases. It was intended to eliminate some of the weaknesses of the purely probabilistic approach described in the previous section. The expert system shell EMYCIN, which evolved out of the MYCIN project, also treats uncertainty in the knowledge base by using the certainty factors approach. In addition to MYCIN, this methodology has been successfully employed in a number of expert systems, including VM, (Fagan, [12]), a ventilator monitoring system and Puff, (Aikin, [2]), an expert system for analyzing pulmonary function medical tests.

Knowledge is expressed in the form of rules and a confidence factor is associated with each rule. For example, here is a modified rule from MYCIN, ([8]):

IF 1) The stain of the organism is gram positive and
2) The morphology of the organism is coccus and
3) The growth conformation of the organism is chains
THEN There is suggestive evidence (0.7) that the identity of the organism is streptococcus.

The value 0.7 represents the confidence factor, CF, associated with the rule. The confidence factor is a value in the interval [-1,+1]. It is not a probability, but it indicates the degree of confidence or belief that the domain expert has on the conclusion being true, given that the condition is true. A certainty factor of -1 indicates total disbelief in the hypothesis given that the evidence is true. A certainty factor of +1 indicates that we are absolutely sure that the hypothesis is true. In contrast to the Bayesian approach, if the rule condition is not true, then no inference is made about the belief or disbelief in the rule hypothesis. In the example given above, the absence of the evidence should not be interpreted as indicating that the organism is not streptococcus with strength of belief 0.3.

The confidence factor assigned to a rule is defined by the formula:

$$CF(h,e) = [MB(h/e) - MD(h/e)] \,/\, (1 - \min(MB(h/e),\ MD(h/e)))$$

where MB(h/e) is the measure of increased belief that h is true given that e is true and MD(h/e) is the measure of increased disbelief that h is true given that evidence e is true. The values of MB and MD range between 0 and 1 and are defined as follows:

$$MB(h/e) = \begin{cases} 1, \text{ if } P(h)=1 \\ [\max\{P(h/e), P(h)\}-P(h)]/(1-P(h)), \text{ otherwise} \end{cases}$$

$$MD(h/e) = \begin{cases} 1, \text{ if } P(h)=0 \\ [P(h)-\min\{P(h/e), P(h)\}]/P(h), \text{ otherwise} \end{cases}$$

In the case where the observed evidence e is in support of the hypothesis h, then P(h/e)>P(h), and therefore the measure of belief, MB, is positive. If the observed evidence decreases the chance that h is true, then MD is positive. If h and e are independent, then (P(h/e)=P(h)), and therefore MB(h/e)=MD(h/e)=0. It can also be seen from the formulas that if CF(h,e)=1 then P(h/e)=1 or in other words, given e then h is certain to be true. Also, if CF(h,e)= -1 then P(¬h/e)=1 or put another way, given e then h is certain to be false.

5.3.2 Updating the CFs of hypotheses for disjunctive evidence

Given two rules, which reach the same hypothesis, h,
$e_1 \Rightarrow h$, $CF(h,e_1)$ and $e_2 \Rightarrow h$, $CF(h,e_2)$
the collective certainty factor assigned to h is given by:

$$CF(h,e_1 \wedge e_2) = \begin{cases} CF(h,e_1) + CF(h,e_2)(1-CF(h,e_1)), \text{ if } CF(h,e_1)>0 \text{ and } CF(h,e_2)>0 \\ CF(h,e_1) + CF(h,e_2)(1+CF(h,e_1)), \text{ if } CF(h,e_1)<0 \text{ and } CF(h,e_2)<0 \\ [CF(h,e_1)+CF(h,e_2)]/(1-\min\{|CF(h,e_1)|, |CF(h,e_2)|\}), \text{ otherwise} \end{cases}$$

Each time a new piece of evidence is introduced, the cumulative value of the confidence factor for each rule hypothesis is updated by the *combination formula* given above. When all evidence has been considered, the hypotheses with the largest certainty factors are suggested as possible solutions.

5.3.2.1 Example on disjunctive updating of CFs

Consider the following predicates and rules:

h_1=high risk investor, h_2= low risk investor, e_1= young, e_2=high income, e_3 = married

Rules:

young $\Rightarrow$ high risk, CF=0.6 or $e_1 \Rightarrow h_1$, CF=0.6
high income $\Rightarrow$ high risk, CF=0.8 or $e_2 \Rightarrow h_1$, CF=0.8
married $\Rightarrow$ high risk, CF= -0.7 or $e_3 \Rightarrow h_1$, CF= -0.7

If both e_1 and e_2 become true, then both rules can fire, and the confidence value of h_1 is updated by the combination formula given above:

$CF(h_1, e_1 \wedge e_2) = 0.6+0.8(1-0.6) = 0.92$

Subsequently, if evidence e_3 becomes available, then the confidence factor on the investor being high risk is updated as:

$CF(h_1, e_1 \wedge e_2 \wedge e_3) = [CF(h_1,e_1 \wedge e_2)+CF(h_1,e_3)] / [1-\min\{|CF(h_1,e_1 \wedge e_2)|, |CF(h_1, e_3)|\}] = [0.92+(-0.7)] / [1-0.7] = 0.22/0.3 = 0.73$

Based on the nonsupporting evidence that the investor is married, the belief, that the investor fits the high risk profile, was significantly reduced from 0.92 to 0.73.

5.3.3 Updating the CFs of hypotheses for conjunctive evidence

In case of uncertainty about the available evidence, or in the case where the available evidence was derived from another rule and it already has a confidence factor assigned to it, the confidence factor on the hypothesis is calculated as follows.

1. In the case of a rule: $e \Rightarrow h$, $CF_{old}(h,e)$ and assuming that the evidence e has a cumulative confidence factor of CF(e), then the confidence on h is updated by:

   $CF_{new}(h,e) = CF(e) \times CF_{old}(h,e)$

2. In the case of a rule of the form:
   $(e_1 \wedge e_2 \wedge .. \wedge e_n) \Rightarrow h$, $CF_{old}(h, e_1 \wedge e_2 \wedge .. \wedge e_n)$.
   Let $CF(e_i)$, i=1,2,...,n be the CF of the premises. Then the overall CF on h is updated by:

   $CF_{new}(h,e_1 \wedge .. \wedge e_n) = CF_{old}(h,e_1 \wedge .. \wedge e_n) \times \min\{CF(e_1),...CF(e_n)\}$.

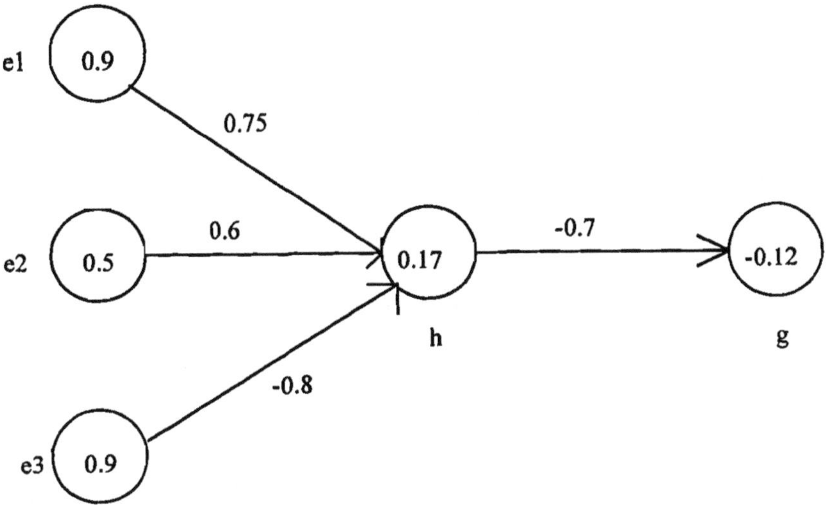

Figure 5.1 Propagation of certainty factors through a rule network.

5.3.3.1 Example of CF propagation in a rule base

Consider the rules:

$e_1 \Rightarrow h$, CF=0.75

$e_2 \Rightarrow h$, CF=0.60
$e_3 \Rightarrow h$, CF= -0.80

where e_1, e_2, e_3, h stand for the predicates in the previous example. In addition consider the rule:

$h \Rightarrow g$, CF= -0.70

where g=invest in blue chip stocks.

A graphical representation of the rule base is given in Figure 5.1.

Assume that the facts e_1 and e_2 are believed to be true with belief values 0.9 and 0.5, respectively. Then after evidence e_1 becomes available, using the chain rule given above:

$CF_{new}(h,e_1)= CF(e_1)\times CF_{old}(h,e_1)=0.9\times 0.75= 0.67$

After evidence e_2 becomes available:

$CF_{new}(h,e_2)= CF(e_2)\times CF_{old}(h,e_2)=0.5\times 0.6= 0.30$

Using the combination formula we get:

$CF(h, e_1 \wedge e_2)= CF(h,e_1)+CF(h,e_2)(1-CF(h,e_1)) = 0.67+0.30(1-0.67) = 0.77$

Next suppose fact e_3 becomes known with belief 0.9 on its validity. Then,

$CF_{new}(h,e_3)= CF(e_3)\times CF_{old}(h,e_3) = 0.9\times(-0.8)= -0.72$

Using the combination formula again for the case of one positive and one negative belief yields:

$CF(h,e_1 \wedge e_2 \wedge e_3)=[CF(h,e_1 \wedge e_2)+CF(h,e_3)]/[1-\min\{|CF(h,e_1 \wedge e_2)|,|CF(h,e_3)|\}]$
$=(0.77+(-0.72))/ (1-0.72)=0.05/0.28= 0.17$

Therefore, based on the evidence, h has a belief of 0.17 on its validity. We can now calculate the belief on g being true:

$CF_{new}(g,h)=CF(h)\times CF_{old}(g,h)= 0.17\times(-0.70) = -0.12$

The negative belief indicates that the investor should not invest in blue chip stocks.

5.3.4 Comparison of the CF and Bayesian approaches

Advantages of the CF approach over the probabilistic approach include the lack of a need for frequency counts or statistical basis in order to supply beliefs for events. The CF approach allows rule representation and quantification of uncertainty at the same time, making easier the implementation of explanation facilities in the system. The simplicity and efficiency with which the certainties are combined is another advantage. As the MYCIN experience showed, experts find it easier to provide CFs for various hypotheses.

On the negative side, the CF approach is not built on a firm theoretical foundation. As a result many weaknesses arise. For example, given that the probability of rain, P(rain), is 0.8 and the probability of rain given a cloudy sky, P(rain/clouds), is 0.9 then the confidence factor CF(rain,clouds)=0.5. It seems to be contradictory that the CF in rain and clouds is less than the probability of rain.

Another disadvantage of the CF approach is that there is an implicit assumption of independence among hypotheses, which of course is not valid in many real world domains.

Certainty factors can be considered as a special case of the Dempster/Shafer theory, ([13]), which is described in Section 5.4.

5.3.5 Certainty factors using Prolog (optional)*

A predicate is written in the form: *predicate(CF, Arg1, ..., Argn)*, where CF is the confidence factor associated with the predicate and Arg_i the predicate arguments. Rules have the form:

```
p(CF1, A1,...,An):- tail1.
...
p(CFn, A1, ..., An):- tailn.
```

where $tail_j$ is a conjunction of predicates.

We give below a possible implementation of a prolog meta-interpreter to propagate and combine the uncertainty factors according to the Certainty Factors approach. The implementation addresses the case where queries have all their argument variables being instantiated. The general case, of finding the overall certainty factors for all answers to queries with unbound variables, is left as an exercise.

The basic prolog predicates are given below. The predicate *and_propagate* evaluates the confidence factor CF_i for predicate p from the overall confidence factors of the predicates in $tail_i$, the conjunctive premise of rule i. The predicate *total* creates a list L, whose elements are the confidence factors CF_i derived by the predicate *and_propagate* for each rule i with head p. The predicate *or_propagate* evaluates the overall confidence factor of predicate p by combining the confidence factors CF_i in list L from each rule which successfully proves the predicate p.

i) The predicate *and_propagate*

Let L be a list whose elements are the overall confidence factors of each predicate in the conjunctive tail of rule: *p(CFand, A1, ..., An):- tail* and *CFrule* be the confidence factor associated to the rule (the confidence in the rule hypothesis being true, given that the predicates in the tail are true with confidence value equal to 1). The predicate *and_propagate* evaluates the overall confidence, *CFand*, in predicate p given that the predicates in the rule tail are believed true with overall confidence less than 1. To this end the formula of Section 5.3.2 is used.

```
and_propagate(L, CFrule, CFand) :- minimum(L,Z), CFand is Z*CFrule.
```

The predicate minimum(L,Z) evaluates the minimum element Z in a list of numbers L and is given by:

```
minimum([X],X).
minimum([X|L],X):-minimum(L,M), X<=M,!.
minimum([X|L],Z):-minimum(L,Z),Z<=X,!.
```

ii) The predicate *or_propagate*

In the case of many rules sharing the same head predicate, the overall confidence factor on the predicate can be evaluated using the *or_propagate* function of Section 5.3.1. The arity of the predicate *or_propagate* is equal to 3. The first argument is a list, L=[CF1,...,CFn], containing the confidence factors derived for predicate p by using the *and_propagate* function on each of the rules having predicate p as their head. The second argument is initialized to 0 and incrementally updated by considering one element of the confidence list L at a time and using the combination formula of Section 5.3.1. The code for the procedure is as follows:

```
or_propagate([ ], CFold, CFold).

or_propagate([C|M], CFold, CFnew):- CFold>=0, C>0,
                                    CF1 is CFold+C*(1-CFold),
                                    or_propagate(M, CF1, CFnew).

or_propagate([C|M],CFold,CFnew):-CFold=<0, C<0,
                                    CF1 is CFold+C*(1+CFold),
                                    or_propagate(M, CF1, CFnew).

or_propagate([C|M],CFold,CFnew):- CFold>0, C<0, K is -C,
```

```
            minimum([K, CFold],W),CF1 is (C+CFold)/(1-W),
            or_propagate(M,CF1, CFnew).

or_propagate([C|M],CFold,CFnew):-CFold<0,C>0, K is -CFold,
            minimum([C,K],W),CF1 is (C+CFold)/(1-W),
            or_propagate(M, CF1, CFnew).
```

iii) The predicate *total*

The predicate *total* is used to search the knowledge base for rules with the same head predicate and create a list containing the *and_propagate* confidence factors from such head predicates. It then applies the function *or_propagate* on the elements of this list to derive the overall confidence factor of the head predicate.

```
total([P|M], Ctotal):- Q=..[P, Cand|M], findall(Cand, call(Q), L),
                        or_combine(L,0,Ctotal).
```

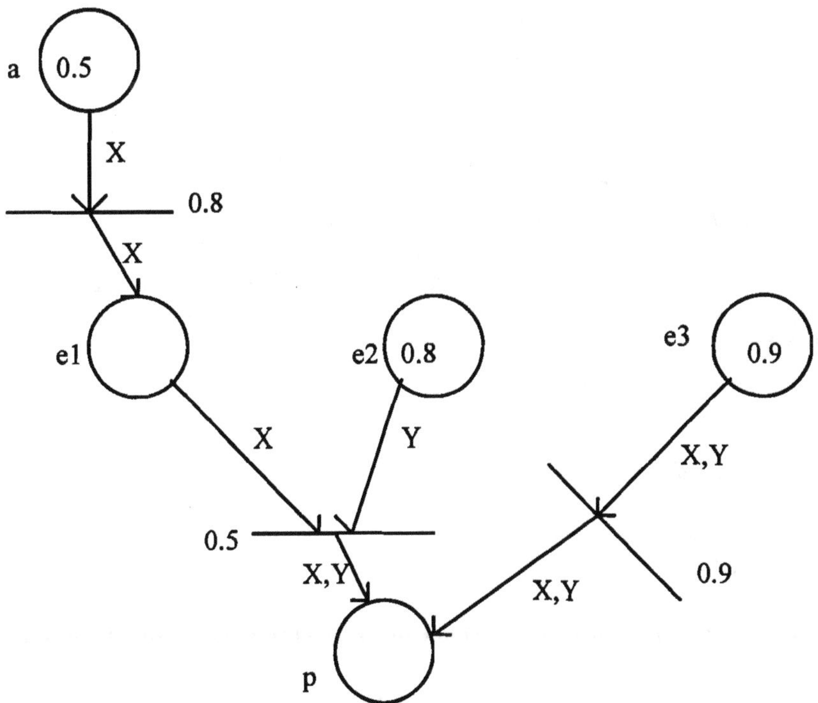

Figure 5.2 Uncertain knowledge represented as a Petri Net.

The procedures given above constitute a general confidence factor evaluation engine. We give an example of how the knowledge base is to be structured to use this engine.

iv) The uncertain knowledge base

Consider the Petri net representation of a knowledge base shown in Figure 5.2. The Petri net model has been modified to accommodate propagation of uncertainty as follows. The confidence factor of a rule is assigned, as a numerical label, to the transition corresponding to the rule. Tokens have been modified and instead of just black and white, they hold the overall belief to the predicate corresponding to the place that contains the token. When a single token goes through a transition the confidence factor the token holds is multiplied by the CF of the transition. When two or more tokens converge to a transition, they merge to a token containing the product of the transition's CF times the minimum of the token weights. If two tokens, containing CFs c1 and c2, are deposited to one place, then they are merged to one token with weight equal to:

c1+c2(1-c1), if c1>0, c2>0
c1+c2(1+c1), if c1<0, c2<0
(c1+c2)/(1-min{|c1|,|c2|}), if c1*c2<0.

The rule base, corresponding to the Petri net knowledge representation shown in Figure 5.2, is structured as follows:

```
p(Cand,X,Y):-total([e1,X], Ctotal1), total([e2,Y], Ctotal2),
                  and_combine([Ctotal1,Ctotal2], 0.5, Cand).
p(Cand,X,Y):-total([e3,X,Y],Ctotal), and_combine([Ctotal,0.9,Cand).
e1(Cand, X):- total([a,X],Ctotal), and_combine([Ctotal],0.8, Cand).
a(0.5,basil).
e3(0.9,basil,christian).
e2(0.8,christian).
```

Queries are asked in the form:

```
?- total([p,basil,christian], Ctotal).
Ctotal=0.85
```

In other words p(basil,Christian) is true with a confidence factor of 0.85.

5.4 DEMPSTER-SHAFER THEORY OF EVIDENCE

The Dempster-Shafer (DS) theory was developed by Dempster, ([11]), and refined and extended by Shafer, ([26]). It was developed to address some of the weaknesses of the probabilistic approach, namely the representation of ignorance, and the unnecessary requirement that the sum of the beliefs to an event and its negation should sum up to one. When asked to assign such measures, human experts do not necessarily maintain the sum of numerical belief assignments of singleton events equal to 1. The DS theory presents an alternative to the Bayesian approach. Both theories communicate degrees of belief, but a probability represents the extent to which a statement is believed to be true, while a belief function represents support. An increase in the likelihood of a hypothesis reduces the likelihood of the compliment, while in DS theory increased support for a hypothesis does not affect the support for the complement. DS theory is more general than the Bayesian method, since belief in a fact and belief in the negation of the fact do not need to sum to 1 (in fact, they could both be equal to 0, if no information is available).

5.4.1 Basic concepts of DS theory

In DS theory, the environment consists of a set of mutually exclusive and exhaustive hypotheses, h_i, $H=\{h_1,h_2,...,h_n\}$, called the *frame of discernment*. Elements of the power set of H, 2^H (set of subsets of H), can be thought of as all possible answers to possible queries. For example, if H={chris, maria, basil} then the answer to the query "who are the females?" is the subset {maria}. The answer to the query "who are the juniors?" is the subset {maria, basil}, etc.

A *mass probability function (or basic probability assignment)*, m, is a function $m:2^H \rightarrow [0,1]$ from the power set of H to the real interval [0,1] such that:

(i) $m(\varnothing)=0$, where $\varnothing$ is the empty event
(ii) $\sum_{A \subseteq H} m(A)=1$

The mass probability assigned to a set A can be interpreted as the portion of the total belief assigned to set A. A subset S of the frame of discernment is a *focal element* iff $m(S) \geq 0$.

The *amount of belief* in a set of hypotheses, A, is denoted by

Bel(A) and is defined as the sum of all mass probabilities of proper subsets of A:

$Bel(A)=\sum_{B\subseteq A} m(B)$.

The mass value assigned to a set A denotes the belief in set A only, while the belief in A, Bel(A), indicates the amount of belief in A or any of its subsets. Since one of the sets in H has to be the answer, we have that Bel(H)=1. The amount of belief assigned to a set of hypotheses can be interpreted as the percentage of the total belief placed on the proposition that the sought after answer is one of the elements in the set.

For example, consider the domain of discernment:

H={wool, rayon, acrylic}

and the process of identifying a fabric as one of these types. A possible mass probability assignment, which describes the knowledge that a given fabric is wool or rayon, is the following:

m({wool, rayon})=0.80
m(H)=0.20
m(A)=0, for every subset A of H other than the ones given above.

The *certainty or belief interval* associated with a subset *A* is the interval: [Bel(A), 1-Bel(A^c)], where A^c denotes the set theoretic complement of A. The quantity 1-Bel(A^c), denoted by Pl(A), is called the *plausibility* of A, and indicates the degree to which the evidence fails to refute A. The plausibility of a subset A is equal to the maximum belief value that can be possibly assigned to A.

The *uncertainty in A or ignorance* is defined by: Pl(A)-Bel(A).

In DS theory, if $X\subseteq Y$ then it is not necessarily true that $m(X)\leq m(Y)$. Also m(H) and $m(X)+m(X^c)$ are not necessarily equal to 1.

The following are easily verified:

$pl(A)\geq Bel(A)$
$Bel(A)+Bel(\neg A)\leq 1$
$Pl(A)+Pl(\neg A)\geq 1$
If $A\subseteq B$, then $Bel(A)\leq Bel(B)$ and $Pl(A)\leq Pl(B)$

In the Bayesian approach, probabilities are represented by a single numerical value. In the DS approach, probabilities are expressed in terms of intervals. It can certainly be argued that it makes more sense to talk about beliefs in terms of intervals instead of requiring exact numerical values. For example, if the certainty range is [0.4, 0.401] then there is little doubt about the measure of belief (it's close to 0.4), while a range of [0.4, 0.9] expresses a lot of uncertainty about the belief in the event.

5.4.2 Updating belief, the combination rule

Assume that m_1 is a mass probability distribution for a frame of discernment, and that a new set of mass beliefs, m_2, is given by the domain expert after additional evidence becomes available, or in another scenario, m_1 and m_2 may be two different mass probabilities assigned to the same frame of discernment by two different experts. In both cases, the updated mass probabilities of the focal elements are given by the following *combination rule*:

$$m_3(C)= [\Sigma_{X\cap Y=C}m_1(X)m_2(Y)]/ [1-\Sigma_{X\cap Y=\varnothing}m_1(X)m_2(Y)]$$

where X,Y,C are subsets of the frame of discernment.

The denominator is a normalization factor, which is introduced in order to maintain the validity of the constraints:

$$m(\varnothing)=0 \text{ and } \Sigma_{A\subseteq H}m(A)=1.$$

The effects of the combination rule are that, as more evidence is collected, belief is partitioned in smaller and smaller subsets, corresponding to more conclusive support for the available hypotheses. When various pieces of evidence implicate different hypothesis, conflict arises. The amount of belief, K, assigned to the empty set $\varnothing$, before normalization (i.e. $K=\Sigma_{X\cap Y=\varnothing}m_1(X)m_2(Y)$), measures the conflict present.

If $\Sigma_{X\cap Y=\varnothing}m_1(X)m_2(Y) =1$, then $m_3(C)$ cannot be defined, and m_1 and m_2 are said to be *contradictory* mass assignments.

5.4.3 An example on the Dempster-Shafer approach

Assume that a choice is to be made between three types of investments.

The investments are denoted by h for high_risk, m for medium_risk and l for low_risk investments. The frame of discernment is:

H={∅, {h}, {m}, {l}, {h,m}, {h,l}, {m,l}, {h,m,l}}.

The mass assigned to each subset of H indicates the domain expert's belief that the set constitutes an answer to the query.

Suppose that the first piece of evidence that becomes available is that the investor is a young adult. Based on this, the financial expert may assign mass values to the focal elements as follows:

$m_1(\{h\})=0$, $m_1(\{h,m\})=0.8$, $m_1(\{l\})=0.1$, $m_1(H)=0.1$.

The last assignment, $m_1(H)=0.1$, is made to the environment of all non-focal sets, and is the part of the total belief that has not been assigned yet.

The second piece of evidence is that the investor is married with children. Based on this, the domain expert assigns the following values:

$m_2(\{m,l\})=0.7$, $m_2(\{m\})=0.2$, $m_2(\{h\})=0.1$, $m_2(H)=0.1$

The total belief associated to set {m, l}, based on the second piece of evidence, is:

$Bel_2(\{m,l\})=m_2(\{m\})+m_2(\{m,l\})=0.2+0.7=0.9$

Table 5.1 Table entries represent the value m1(X)m2(Y), where X and Y are sets corresponding to row and column.

	m1({h,m})=0.8	m1({l})=0.1	m1(H)=0.1
m2({m,l})=0.7	0.7×0.8=0.56	0.07	0.07
m2({m})=0.2	0.16	0.02	0.02
m2(H)=0.1	0.08	0.01	0.01

The normalization factor N is equal to:

$N=1-\sum_{X\cap Y=\varnothing}m_1(X)m_2(Y)= 1-0.02= 0.98$

The updated beliefs are then calculated as:

$m_3(\{m\})= (1/N)\sum_{X\cap Y=\{m\}}m_1(X)\times m_2(Y) = (0.56+0.16+0.02)/0.98 = 0.7551$
$m_3(\{l\}) = (1/N)\sum_{X\cap Y=\{l\}}m_1(X)\times m_2(Y)= (0.07+0.01)/0.98 = 0.0816$
$m_3(\{h,m\}) =(1/N)\sum_{X\cap Y=\{h,m\}}m_1(X)\times m_2(Y) = 0.08/0.98 = 0.0816$
$m_3(\{m,l\}) = (1/N)\sum_{X\cap Y=\{m,l\}}m_1(X)\times m_2(Y) = 0.07/0.98 = 0.0714$
$m_3(H) = (1/N)\sum_{X\cap Y=H}m_1(X)\times m_2(Y) = 0.01/0.98 = 0.0102$

The corresponding beliefs and plausibilities are:

$Bel(\{m\})= \sum_{X\subseteq\{m\}}m_3(X)= 0.74/0.98 = 0.7551$
$Pl(\{m\}) = 1- \overline{Bel}(\{m\}^c) = 1- Bel(\{h,l\}) = 1-0.08/0.98 = 0.9184$
$Bel(\{l\}) = 0.08/0.98 = 0.0816$
$Pl(\{l\})=1-Bel(\{l\}^c)=1-Bel(\{h,m\})=1-(Bel(\{h,m\})+Bel(\{m\})=$
$1-(0.08+0.74)/0.98 = 0.1633$
$Bel(\{h,m\}) = m_3(\{h,m\})+m_3(\{m\}) = (0.08+0.74)/0.98 = 0.8367$
$Pl(\{h,m\}) = 1 - Bel(\{h,m\}^c) =1-Bel(\{l\})= 1-0.08/0.98 = 0.9184$
$Bel(\{m,l\}) = m_3(\{m,l\})+m_3(\{m\})+m_3(\{l\}) = (0.07+0.74+0.08)/0.98=$
0.9082
$Pl(\{m,l\}) = 1-Bel(\{m,l\}^c) = 1-Bel(\{h\}) = 1-0.163=0.837$

The corresponding confidence intervals are:

for {m}: [0.7551, 0.9184]
for {l}: [0.0816, 0.1633]
for {h,m}: [0.8367, 0.9184]
for {m,l}: [0.9082, 0.9234]

Consequently, given the available evidence, the investor should be classified as medium risk, ({m}).

5.4.4 Comparisons

Shafer, ([26]), has shown that when the DS combination rule is used to combine Bayesian basic probability assignments, the result is equivalent to Bayes's theorem. Therefore probability theory is a special case of belief functions.

A difficulty with the DS theory is that the theory does not specify how the mass probabilities are to be computed or how the results (belief intervals) are to be interpreted, (Zadeh, [30]). In addition, obviously incorrect conclusions can be reached in certain cases.

For example, consider two physicians A and B whose beliefs in the patient's disease are:

m_A(flu)=0.99, m_A(pneumonia)=0.01 and
m_B(cold)=0.99, m_B(pneumonia)=0.01.

The application of the combination rule of DS theory will give that the belief in pneumonia is 1, clearly a contradiction to the available belief of the experts.

The DS approach allows beliefs to be assigned to sets of events rather than to just a finite number of mutually exclusive events, as was the case in the Bayesian approach. The Bayesian approach represents ignorance by equally distributing the likelihoods among all single hypotheses. This makes the representation of ignorance context dependent. For example, if no information is available about identifying an incoming flying target and there are 100 possible targets, probability theory will assign a probability of 0.01 to each of the targets, while in DS theory a mass function value of 1 is assigned to the frame of discernment H and 0 to all other subsets.

The following example shows the weakness of the Bayesian approach, which results from forcing numerical values as probabilities of single events.

Suppose that in the database of investors, due to the large size of the database, only 65% of the investors are sampled. Assume that all investors picked in the sample happen to be young adults and that all of them have invested in high risk portfolios. The real probability that a young adult investor invests in a high risk portfolio is of course not 1, as the sample would have us believe, but is between 0.65 and 1.0, since it is possible that either all or none of the remaining young adult investors invested in high risk portfolios. The probability that a young adult will not invest in high risk portfolio is then between 0.0 and 0.65. In DS theory, there is some direct evidence to support high risk investment (65%) while the other 35% is assigned as a belief or commitment to ignorance, and is not distributed equally to the rest of the events.

In contrast, in the Bayesian approach the remaining investment types of moderate and low risk will be assigned probabilities of 0.35/3=0.116 each under the Bayesian principle of indifference. In effect, the Bayesian approach is forcing numerical values on events to estimate their likelihood when in fact the available evidence does not support them. Suppose that a second piece of evidence is observed supporting the high or moderate risk hypothesis. Then the probability that the investor should invest in high risk portfolios would become:
0.65/(0.65+0.116)=0.85, which is unreasonably high for the evidence available.

5.5 FUZZY SETS AND FUZZY LOGIC

5.5.1 Introduction

Knowledge is not always expressed in boolean, crisp form. Most often, reasoning takes place using information in imprecise or vague form. For example, knowing that

Computer chess programs usually beat players with low ratings

and that

Usually the players that participate in junior tournaments have low ratings

then we can conclude that

Computer chess programs usually beat junior tournament players.

Fuzzy logic is an attempt to do away with the boolean paradigm and deal with vagueness in the knowledge. Fuzzy set theory and fuzzy logic were developed by Zadeh, ([33]), extending Lukasiewicz's n-valued logics. In n-valued logic, the set of truth values was extended from just 0 or 1 (true or false) to values in the truth set $T_n=\{0, 1/(n-1), 2/(n-1), ..., (n-2)/(n-1), 1\}$. For example, in 3-valued logic the truth values allowed are 0, 1/2 and 1, corresponding to false, unknown and true. The standard Lukasiewicz logic L1, is an infinite valued logic, where the truth values can span the whole interval [0,1] and corresponds to Zadeh's fuzzy set theory.

Experts often incorporate in their reasoning process qualitative concepts such as *high, low, hot,* etc., and quantifiers like *very, a little, usually, sometimes*, etc. The inherent vagueness of these terms is naturally implemented by fuzzy logic and reasoning with such terms can take place using fuzzy inference. In such domains where very little or no a priori knowledge is available, the probabilistic approach is rendered ineffective.

Some of the major applications of fuzzy logic to expert system development include its use to:

a) Control trains in Japan using fuzzy controllers, (Miyamoto, Yasunobu, [21]).
b) Cement Kiln controller, (Mamdani, Gaines, [19]).
c) FLOPS is a fuzzy ES rule based shell, (Buckley et al., [9]).
d) Z-II is a fuzzy ES shell used in medical diagnosis and risk analysis, (Leung and Lam, [17]).
e) Fuzzy logic has been applied in video camera technology for automatic focusing, automatic exposure, image stabilization and white balancing.
f) Fuzzy logic has been applied in automobiles for cruise control, brake and fuel injection systems.
g) Fuzzy algorithms have been applied for video and audio data compression (HDTV).

5.5.2 Fuzzy set theory

In traditional set theory, a set A can be defined in terms of a function, $\mu_A:U\rightarrow\{0,1\}$, from the universe of discourse (universal set) U to the

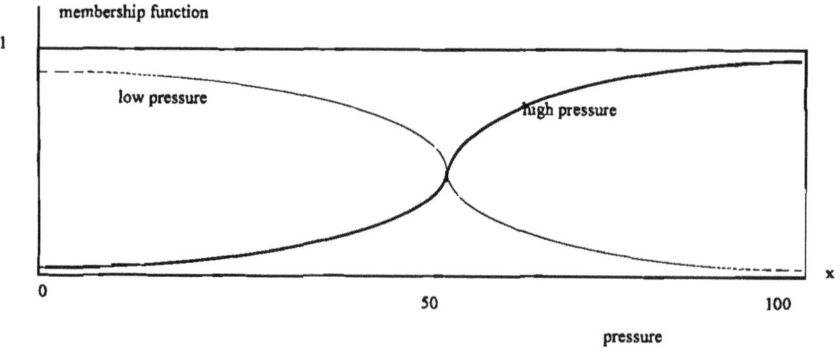

Figure 5.3 Two fuzzy sets: high pressure and low pressure.

discrete set {0,1}, where $\mu_A(x)=1$, if x belongs to the set A, 0 otherwise. These sets are called *crisp sets* and membership of an element to a crisp set has the truth value of either true or false.

A fuzzy set generalizes the notion of crisp sets. A *fuzzy set* A is defined by a function, $\mu_A:U\rightarrow[0,1]$, from the universal set U to the continuous interval [0,1], called the *membership function*. The domain of the membership function of a fuzzy set can be discrete or continuous. The image of an element x, $\mu_A(x)$, is a real number in the closed interval [0,1] which indicates the *strength* of the membership of x to the set A. It should be noted that the strength of membership of an element in a set is not the probability that the element belongs to the set. The membership degrees of all the elements in U do not have to sum up to one.

For example, consider the concept of *tall* and let the universal set be all faculty and students at BU. In boolean logic, someone is either tall or not tall. There is no middle ground. To represent tall as a crisp set, a threshold value would have to be chosen, and all elements of the universal set, whose height is greater than the threshold value, would be classified as *tall* and less than the threshold value as *not tall*. A height within one tenth of an inch off the threshold and a height of two feet off the threshold would both be classified as "not tall". In contrast, in fuzzy

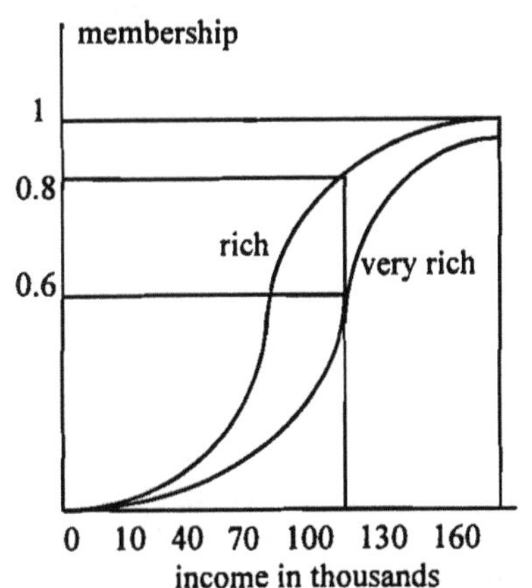

fuzzy set rich and quantified set very rich

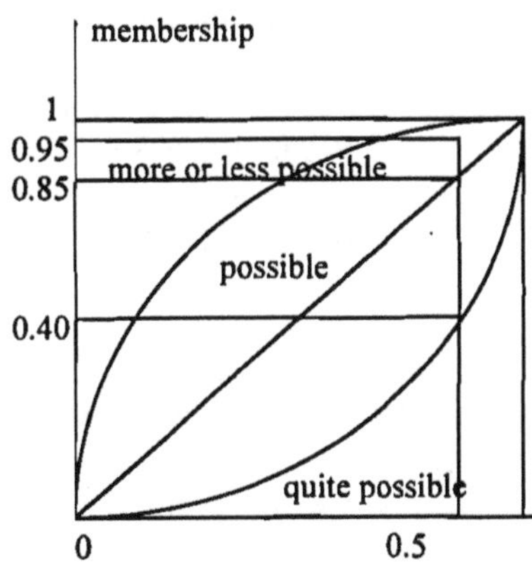

Linguistic truth values possible, more or less possible, quite possible

Figure 5.4 Fuzzy linguistic terms.

set theory a membership value is assigned to each height, indicating the amount of confidence in being tall. A height of 6.6' may be assigned a membership value of 1.0 (he is definitely tall), while a height of 5.11' may be assigned a membership value $\mu_{tall}(5.11)= 0.80$ and a height of 5.2' may be assigned $\mu_{tall}(5.2)=0.02$.

A graphical representation of two fuzzy sets, *high* and *low*, is depicted in Figure 5.3.

It should be noted that constructing an appropriate membership functions is one of the main difficulties in using fuzzy logic and fuzzy inference. The choice of membership function can affect the system performance. The methods, which have been proposed so far, are experimental and ad hoc.

The *support* of a fuzzy set A, *sup(A)*, is defined to be the set of elements from the universal set whose degree of membership is positive: $sup(A)=\{u \in U/\mu_A(u)>0\}$.

A fuzzy set A of finite support is usually denoted by $A=\sum_{i=1,n}\mu_i/x_i$, where $sup(A)=\{x_1, x_2, ..., x_n\}$ and μ_i is the membership degree of x_i.

Consider for example the fuzzy sets "high_pressure" and "low_pressure". If the domain is discrete, they can be defined as follows:

high_pressure=0.0/3.0+0.0/3.5+0.0/4.0+0.0/4.5+0.1/5.0+0.4/5.5+
0.9/6.0+1.0/6.5
low_pressure=1.0/3.0+1.0/3.5+1.0/4.0+0.8/4.5+0.5/5.0+0.1/5.5+
0.0/6.0+0.0/6.5

Linguistic terms, fuzzy quantifiers.

The certainty degrees of statements can be represented by *linguistic terms* like *possible*, *rather possible*, *very possible*, etc. Linguistic terms are represented as functions from [0,1] to [0,1]. For example, we can say that : *"It is quite possible that Basil is rich"*. Here rich is a fuzzy set and quite possible is a linguistic term which assigns a truth value to the statement that Basil is rich. A graphical representation is given in Figure 5.4. The fuzzy set rich is defined in terms of yearly income. According to the graph if Basil's income is 250,000 dollars then his degree of membership to the set rich is 0.85. From the graphic definition of *quite possible* and *possible*, we then get that the degree of truth of the statement *"It is quite possible that Basil is rich"* is 0.8 while the degree of truth in the statement *"It is possible that Basil is rich"* is 0.9.

The membership of an element in a fuzzy set is often quantified by such quantifiers as: extremely, very, more or less, quite etc. Membership functions can be defined for such quantifiers to express their intended meaning. For example, if A is a fuzzy term then the membership function of the quantifiers *very*, *more_or_less* and *not* are defined by:

$\mu_{very(A)}(x)=\mu_A{}^2(x-\delta)$
$\mu_{more\_or\_less(A)}(x)=\sqrt{\mu_A}(x+\delta)$
$\mu_{not(A)}(x)=1-\mu_A(x)$

where δ is a *focus shifting* constant.

An ordinary fuzzy set is sometimes called a *type 1* fuzzy set. A *type 2* fuzzy set is a fuzzy set whose membership degrees are themselves type 1 fuzzy sets. For example, "happy" can be viewed as a type 2 fuzzy set whose membership degrees assume the values "somewhat", "very", "extremely", etc., which are themselves fuzzy sets. Inductively, fuzzy sets of type n can be defined for all $n \geq 1$.

Operations on fuzzy sets.

There are many alternate definitions in the literature for the set operations of union, intersection, etc. for fuzzy sets. The most commonly used are the following:

$\mu_{A\cap B}(x)= \min(\mu_A(x), \mu_B(x))$
$\mu_{A\cup B}(x)= \max(\mu_A(x), \mu_B(x))$
$\mu_{A\Rightarrow B}(x)=\mu_{\neg A\vee B}(x)= \max(1-\mu_A(x), \mu_B(x))$
$\mu_{A\times B}(u,v)= \min(\mu_A(u), \mu_B(v))$, where AxB is the cartesian product of A and B.
$A\subseteq B$ iff $\mu_A(x)\leq\mu_B(x), \forall x\in U$

It should be noted that the fuzzy operators satisfy all boolean properties except $A\cup A^c=U$ and $A\cap A^c=\emptyset$.

Other classes of membership functions have also been explored, for example:

$\mu_{not(A)}(x)=(1-\mu_A(x))/(1+\lambda\mu_A(x))$, $\lambda\in[-1,+\infty)$ (Sugeno class)

$\mu_{not(A)}(x)=(1-\mu_A(x)^w)^{1/w}$, $w\in(0,+\infty)$ (Yager class)
$\mu_{A\cup B}(x)= \min[1, (\mu_A(x)^w+\mu_B(x)^w)^{1/w}$
$\mu_{A\cap B}(x)=1-\min[1, [(1-\mu_A(x))^w+(1-\mu_B(x))^w]^{1/w}]$, $w\in(0, +\infty)$

A relation between two fuzzy sets is defined as follows:

A *fuzzy relation* R from a crisp set U to a crisp set V, is a fuzzy subset of U×V, $R=\{(u,v)/m_R(u,v)$, where $u\in U$, $v\in V\}$, where m is the membership function.

A *fuzzy relation R* from a *fuzzy set A* to a *fuzzy set B* is the following fuzzy subset of U×V: $R=\{(u,v)/m(u,v)$ with $m(u,v)\leq\mu_A(u)$, $m(u,v)\leq\mu_B(v)$, $u\in U$, $v\in V\}$

5.5.3 Fuzzy implications

The analysis presented in this section is based on a paper by Bouchon-Meunier, ([7]).

A fuzzy implication is of the form:

IF X is A THEN Y is B

where A,B are fuzzy sets defined on a universal sets U_X and U_Y respectively. For example, the rule

IF temperature is low THEN oil viscosity is high

is a fuzzy rule.

The definition of the truth table of a fuzzy implication should be a generalization of the boolean case. In other words if the condition "X is A" is satisfied with a degree of membership $\mu_A(X)=1$ (absolutely true) and "Y is B" with a membership degree of $\mu_B(Y)=0$ (absolutely false), then the truth value of the implication $\mu_{A\Rightarrow B}(X,Y)$ should be equal to 0. If $\mu_A(X)=0$ or ($\mu_A(X)=1$ and $\mu_B(Y)=1$) then the membership value of the implication $\mu_{A\Rightarrow B}(X,Y)$ should be equal to 1.

Many alternate definitions of fuzzy implication have been given in the literature which satisfy the conditions given above and extend the boolean implication, ([7]). We give below some of these possible membership functions.

$\mu^1{}_{A\Rightarrow B}(X,Y)=1-\mu_A(X)+\mu_A(X)\mu_B(Y)$
(Reichenbach)
$\mu^2{}_{A\Rightarrow B}(X,Y)= \max[1-\mu_A(X), \min(\mu_A(X),\mu_B(Y))]$ (Willmott)
$\mu^3{}_{A\Rightarrow B}(X,Y)= \min(\mu_A(X), \mu_B(Y))$ (Mamdani)
$\mu^4{}_{A\Rightarrow B}(X,Y)= 1$, if $\mu_A(X)\leq \mu_B(Y)$ and 0 otherwise (Rescher-Gaines)
$\mu^5{}_{A\Rightarrow B}(X,Y)= \max(1-\mu_A(X), \mu_B(Y))$ (Klene-Dienes)
$\mu^6{}_{A\Rightarrow B}(X,Y)= 1$, if $\mu_A(X)\leq \mu_B(Y)$, or $\mu_B(Y)$ otherwise (Brouwer-Godel)
$\mu^7{}_{A\Rightarrow B}(X,Y)= \min(\mu_A(X)/\mu_B(Y), 1)$, if $\mu_B(Y)\neq 0$,
or =1, otherwise (Gorgen)
$\mu^8{}_{A\Rightarrow B}(X,Y)= \min(1-\mu_A(X)+\mu_B(Y), 1)$

Observe that:

$\mu^8{}_{A\Rightarrow B}(X,Y)=0$ iff $\mu_A(X)=1$ and $\mu_B(Y)=0$
and
$\mu^8{}_{A\Rightarrow B}(X,Y)=1$ iff $\mu_A(X)\leq \mu_B(Y)$.

In other words μ^8 generalizes the boolean implication, but it can also assume the value 1 for cases that the boolean implication does not cover.

A weakness of all of the given definitions of implications except μ^3, is the following:

Consider the rule: IF X is A THEN Y is B and assume that the fact X is A' holds for some fuzzy set, A', different than A. Using the given formulas, it is possible that a $Y\in U_Y$ exists such that $\mu_{B'}(Y)$ is different than 0 even though $\mu_B(Y)=0$. As a result, something that the rule says is not possible, may become possible.

These formulas can be generalized for fuzzy rules of conjunctive conditions:

IF X_1 is A_1 and ... and X_n is A_n THEN Y is B, where A_i's, B are fuzzy sets.

For example μ^8 becomes in that case:

$$\mu^8{}_{A1\wedge..\wedge An\Rightarrow B}(X_1,\ldots,X_n,Y) =\min[1-\min(\mu_{A1}(X_1),\ldots,\mu_{An}(X_n))+\mu_B(Y),1] \quad (6.1)$$

Partial match of fuzzy rule conditions

Given the rule

IF (temperature is low) THEN (oil_viscosity is high)

and a fact such as *the temperature is 10° F* or *the temperature is very low*, the rule will not fire if a total match is expected between facts and rule conditions. A generalized rule firing technique has been introduced, (Zadeh, [29]) to deal with this case.

Given the rule:

IF X is A THEN Y is B, and a fact *X is A'*,

a conclusion *Y is B'* can be reached. The membership function of B' should be defined in such a way that if A=A' then $\mu_B=\mu_{B'}$. This is done as follows:

Given the membership function of A', $\mu_{A'}$, the membership of the fuzzy set B' is defined by:

$$\mu_{B'}(Y) = sup_{X \in Ux}[max_X(\mu_{A'}(X)+\mu^i_{A \Rightarrow B}(X,Y)-1,\ 0)]. \qquad (6.2)$$

For i=1,2,3,5,6,7,8 we have that the desired property, (if A=A', then $\mu_{B'}=\mu_B$), holds.

Alternate formulas preserving this property, that if A=A' then $\mu_{B'}=\mu_B$, have also been proposed. For example,

$$\mu_{B'}(Y) = max_X\,(min(\mu_{A'}(X),\ \mu^i_{A \Rightarrow B}(X,Y)))$$

which preserves this property for i=3 or 6.

The partial match of conditions formula can now be generalized for rules with conjunctive hypotheses as follows:

Given a rule of the form

IF $(X_1$ *is* A_1 *and ... and* X_n *is* $A_n)$ *THEN Y is B*

and the facts X_1 *is* A_1', ..., X_n *is* A_n', the membership function of the conclusion is given by:

$$\mu_{B'}(Y)=\max_{x1,\ldots,xn}[\min(\mu_{A1'}(X_1), \ldots, \mu_{An'}(X_n), \mu^{i}_{A1,\ldots,An\Rightarrow B}(X_1,\ldots,X_n,Y))] \quad (6.3)$$

5.5.4 Rules with relational conditions

Rules with fuzzy relational conditions can have their membership function evaluated by converting them to fuzzy rules of the form given in the previous section ([31]).

Consider the rule :

IF (interior_temperature is equal to exterior_temperature) THEN lower the thermostat.

or in general a rule of the form:

IF (X Related_to Y) THEN Z is B

where *Related_to* is a fuzzy relation like "close to", "equal to", etc., and B is a fuzzy predicate.

This rule can be rewritten as:

IF (X,Y) is in relation Related_to THEN Z is B,

where Related_to$\subseteq U_X \times U_Y$ is the relationship viewed as a fuzzy set. Then

$$\mu^{g}_{rule}(X,Y,Z)=\min(1,\ 1-\mu_{related\_to}(X,Y)+\mu_B(Z))$$

If the additional facts *X is D* and *Y is E* are known then by formula (6.2):

$$\mu_{rule}(X,Y,Z)=\min(\mu_D(X),\ \mu_E(Y),\ \min(1,\ 1-\mu_{related\_to}(X,Y)+\mu_B(Z)))$$

and finally by projection:

$$\mu_{B}'(Z)=\max_{X,Y}[\min(\mu_D(X),\ \mu_E(Y),\ \min(1,\ 1-\mu_{related\_to}(X,Y)+\mu_B(Z)))]$$

We demonstrate the usage of this formula by an example.

5.5.5 Example on fuzzy decision making

Consider the rule:

IF (outside_temperature X is low and inside_temperature Y is close to X) THEN (turn the thermostat temperature Z to high).

and the facts:

outside_temperature X=20° F, inside_temperature Y=30° F.

Let us assume that the membership functions of *low(X)* and *high(X)* are given by:

low(X)=1.0/10°F+0.7/20°F+0.6/30°F+0.2/40°F
high(X)=0.0/10°F+0.2/20°F+0.4/30°F+0.8/40°F

and the membership function for the relation *close_to(X,Y)* is given by the following matrix C, where $C(i,j)=\mu_{close\_to}(i,j)$.

C=

	10°F	20°F	30°F	40°F
10°F	1.0	0.8	0.5	0.0
20°F	0.8	1.0	0.6	0.2
30°F	0.5	0.6	1.0	0.9
40°F	0.0	0.2	0.9	1.0

The given relational rule and facts can be rewritten as:

R1: IF (X is low and (X,Y)∈Close) THEN Z is high
F1: X is 20
F2: Y is 30

The facts will have to be *fuzzified.* One way to convert precise values into fuzzy sets is as follows:

$\mu_{F1}(X)$= 1, if X=20 and 0, otherwise
$\mu_{F2}(Y)$= 1, if Y=30 and 0 otherwise.

According to (6.1):

$$\mu^{8}_{R1}(X,Y,Z)= \min [\ 1,\ 1-\min(\mu_{low}(X),\ \mu_{close}(X,Y))+\mu_{high}(Z)]$$

Therefore according to (6.3):

$\mu_{high'}$ (Z)= max $_{X,Y}$[min(μ_{F1} (X), μ_{F2} (Y), $\mu^{8}{}_{R1}$ (X,Y,Z))]=
max$_{X,Y}$[min(μ_{F1} (X),μ_{F2} (Y),min[1,1-min(μ_{low} (X), μ_{close} (X,Y))+μ_{high} (Z)])]

So:

$\mu_{high'}$ (Z) = $\mu^{8}{}_{R1}$ (20,30,Z)= min [1, 1-min(μ_{low}(20),μ_{close}(20,30))+μ_{high} (Z)]=
min(1,0.4+μ_{high} (Z)) = 0.4/10°F +0.6/20°F +0.8/30°F + 1.0/40°F

This formula gives the membership function of the thermostat temperature Z in the fuzzy set *high'* given facts F1 and F2.

Next assume that a second rule was available:

IF (outside temperature is high and inside temperature is close to the outside temperature)
THEN (set the thermostat temperature Z to low).

Given the facts F1 and F2, calculating the membership function of Z as above yields:

$\mu_{low'}$ (Z)=...=min(1, 0.8+μ_{low} (Z))=1.0/10°F+1.0/20°F+1.0/30°F+1.0/40°F

Defuzzification

When a precise answer is desired out of a fuzzy knowledge base, the process of *defuzzification* is applied to convert fuzzy conclusions to concrete results. One way to achieve this is by calculating a weighted average of the best choices from each fuzzy set. From the membership function, $\mu_{high'}$ (Z), the temperature 40 is the centroid among the temperatures of largest truth value, and for the membership function, $\mu_{low'}$(Z), the centroid is 25°F. The corresponding strengths are 1.0 and 1.0. The area bounded by the graph of the membership function for high'(Z) and the horizontal line at the membership value of the centroid is equal to 12.75 and the corresponding area in the graph for low'(Z) is equal to 30.0. The weighted average is: (40×12.75+25×30.00)/(12.75+30.0)=29.4 and this will be the defuzzified value to which to set the thermostat on.

Note: If the thermostat setting Z appears in the condition of another rule, then the newly derived membership functions $\mu_{high'}$(Z) and $\mu_{low'}$(Z) will be used in the inference chain.

5.5.6 Fuzzy logic in Prolog (optional)*

Fuzzy sets can be represented in Prolog as follows:

fuzzyset(Type, FuzzyTerm, [[X1,Y1], ..., [Xn,Yn]]).

"FuzzyTerm" is the name of the fuzzy set and "Type" is a variable which ranges over the domain of definition of the fuzzy set. The third argument is a list called the *membership list* of the fuzzy set Fuzzyterm. In this list the X_i's are elements in the universal set and the Y_i's their corresponding membership degrees.

For example, the membership function of the fuzzy set "high" of the previous section can be represented as the Prolog fact:

fuzzyset(temperature, high, [10:0.0, 20:0.2, 30:0.4, 40:0.8])

The membership function of fuzzy quantifiers such as "very" can be defined by rules of the following type:

% if [X,M] is in the membership list of A, then [X, M*M] is in the %membership list of very(A)

```
fuzzyset(Type, very(A), Membership_list_of_very_A):-
        fuzzyset(Type,A, Membership_list_of_A),findall([X,Y],
        (member([X,M],Membership_list_of_A), Y is M*M),
        Membership_list_of_very_A).
```

Prolog rules to define other quantifiers like "more or less", "not", etc. can be similarly constructed.

Facts are "fuzzified" and denoted by the following clause format:

fuzzyset(Type, v, [v:1.0]).

In other words, if a fuzzy variable is given a crisp value v, then the membership function is 1.0 at that value and 0.0 elsewhere. For example, the fact that the temperature is 10°F corresponds to the clause:

```
fuzzyset(temperature, 10, [10,1.0]).
```

As an example, the fuzzy rule base given above can be expressed in Prolog as follows:

```
is(thermostat, Z, high):-is(outsideTemperature,O,low),
                          is(insideTemperature, I, close(O)).
fuzzyset(outsideTemperature, 10,[10,1.0]).
fuzzyset(insideTemperature, 20,[20.1.0]).
fuzzyset(Temperature,close(Y),[[X6,0.0],[X5,0.2],[X4,0.7],[X,1.0],
[X1,0.7], [X2,0.2], [X3,0.0]]) :- X is Y,X1 is Y+10,X2 is Y+20,
                          X3 is Y+30,X4 is Y-10,X5 is Y-20, X6 is Y-30.
fuzzyset(Temperature, low, [[10,1.0], [20,0.7], [30,0.6], [40,0.2]]).
fuzzyset(Temperature, high, [[10,0.0], [20,0.2], [30,0.4], [40,0.8]]).

% eval(is(T,X,H),C) gives the truth value H that fuzzy set T is X  as an
% answer to the query ?is(T,X,H). If there is a fact giving the exact value
% of T, then H is equal to 1.0

eval(is(T,X,H), 1.0):-fuzzyset(T,X,[X,1.0]), H=[X,1.0].

% If the exact value of T is given in a fact and H is a fuzzy set, "eval" %
returns C, the membership degree of the  value for T in the fuzzy set H

eval(is(T,X,H),C):-fuzzyset(T,H,Mlist),member([X,M],Mlist), C is M, !.

% This rule is for the case where there is no fact giving the membership
% of X in the fuzzy set T, but there is a rule with head unifiable to the
% query "?is(T,X,H)"

eval(is(T,X,H),C):-
            clause(is(T,X,H),B),eval(B,BC),fuzzyset(T,H,Mlist),
            member([X,M], Mlist),S is 1-BC+M, min(1,S,C).

% the following rule evaluates the fuzzy truth value in case the body of
%the rule contains more than one predicate

eval((B1,B2),C):-eval(B1,C1),eval(B2,C2),min(C1,C2,C).
member(X,[X]).
member(X,[Y|Z]):-member(X,Z),!.
min(X,Y,M):-X=<Y,M=X.
min(X,Y,M):-Y=<X,M=Y.
```

The fuzzy knowledge base can then be queried to derive the membership values of different thermostat settings:

```
?eval(is(thermostat, Z, high,), C).
Z=10, C=0.4
Z=20, C=0.6
Z=30, C=0.8
Z=40, C=1.0
```

For a more detailed and the complete fuzzy inference engine, the reader is referred to T. Van Le, ([30]), from where much of the above discussion was inspired.

5.5.7 Fuzzy logic programming environments

A number of systems have been built which support fuzzy logic programming. *FProlog*, [20], is a fuzzy Prolog interpreter in which the updating and propagation of the fuzzy membership values through the rules is done automatically by the inference engine. Backtracking takes place when the truth value of a predicate falls below a threshold value. The negation operator *not* assigns the truth value of 1-p to the clause *not(predicate)* where p is the truth value of *predicate*.

Other systems which support fuzzy logic programming include FRIL, [5], the expert system shell Z-II, [16], and the *support logic programming* environment, [6]. In support logic programming, each clause is accompanied by a pair of numbers (the *support pair*) representing the largest possible and the smallest possible amount of belief in the clause being true.

5.6 BAYESIAN BELIEF NETWORKS

5.6.1 Weaknesses of probabilistic approaches

One weakness of the previous approaches is their inability to simultaneously handle *predictive* and *diagnostic* inferences. For example, consider the rules:

IF rate of inflation is high THEN the interest rate is up
(predictive inference),

and

IF the interest rate is up THEN the rate of inflation is probably high (diagnostic inference).

Assume that both rules are contained in the knowledge base and the CF approach is used. Let the confidence factor of the first rule be 0.9 and of the second 0.4. Given the fact "inflation is high", the fact "interest rate up" can be deduced with certainty 0.9, which in turn activates the second rule, recalculating the confidence in high inflation to 0.36 (doesn't make sense). Repetitive tracing through the rule cycle would eventually get the confidence to both high inflation and high interest rate to be close to 0. To avoid this difficulty, such cycles in the rule base are not allowed under the certainty factors approach. This deprives the system from either its predictive or diagnostic abilities.

Another weakness is demonstrated by the following example. Consider the rules,

IF waves are big THEN it is windy, CF=0.8 and
IF a ship goes by THEN waves are big, CF=1.0.

Given the evidence that a ship went by, using transitivity we would still conclude that it is windy with a CF of 0.8. This should not happen; the presence of the ship should reduce our belief in wind. To achieve this would require another rule directly relating ship and wind, or the passing of a ship has to be introduced as the exception to the first rule. The concepts of ship and wind are only negatively related in the presence of big waves. As a result, the knowledge engineer would probably neglect to include any exception rules or any negative correlation rules. This would allow the erroneous conclusion of wind, given the passage of a ship.

Another problem is the locality of CF propagation. If evidence e_1, e_2, e_3 all can imply h and they are all available with a certain CF, then the confidence in h is increased from each input. This can result in an overestimate of the certainty assigned to h, in the case where the e_i's were derived from a single condition. Consider for example the rule base:

$a \Rightarrow c$, CF=1.0
$a \Rightarrow d$, CF=1.0
$c \Rightarrow h$, CF=0.6

$d \Rightarrow h$, CF=0.6

Given the fact a, we conclude h with certainty factor CF(h)=0.6+0.6(1-0.6)=0.84. This is clearly an overestimate.

Given n boolean variables, it requires 2^n probabilities to specify the joint probability distribution $P(X_1,\ldots,X_n)$. To calculate the conditional probability $P(X_i/X_j)$ would require finding the marginal probabilities $P(X_i \wedge X_j)$ and $P(X_j)$, a task in itself involving summing up 2^{n-1} probabilities representing combinations of values for the remaining variables. To reduce the number of required probabilities and make the problem computationally tractable, a simplified form of Bayes' rule was derived in Section 5.2, under the assumption that all the evidence was pairwise independent. In that case, $P(e_1,..,e_n/c)=P(e_1/c)...P(e_n/c)$ requiring only 4n probabilities to be supplied, ($\forall i=1,\ldots,n$ we need to know: $P(e_i/c)$, $P(e_i/\neg c)$, $P(\neg e_i/c)$, and $P(\neg e_i/\neg c)$). Even though this assumption reduced the exponential amount of required data to linear, semantically it is an undesirable assumption and rarely valid in practice.

A Bayesian belief network is a well accepted method whose usage has recently become widespread and which addresses the weaknesses of older numerical methods , ([1, 4, 22, 23]).

5.6.2 Belief networks, definitions

Bayesian or belief networks, (Pearl, [24]), are an extension of the probabilistic approach. They address some of the issues mentioned above and do not require the assumption of conditional independence among all pieces of evidence. A *bayesian belief network* is a directed acyclic graph (DAG) with the nodes representing statistical variables and the arcs indicating causal dependencies between the variables. The strengths of the influences between nodes X and Y are represented by conditional probabilities. An arc from node X to node Y indicates that X causally influences Y. Observing evidence X provides causal support for Y and observing Y provides diagnostic support for X. In addition to direct causal relationships, the belief network representation facilitates the discovery of induced dependencies among causes.

As an example, consider the following knowledge base:

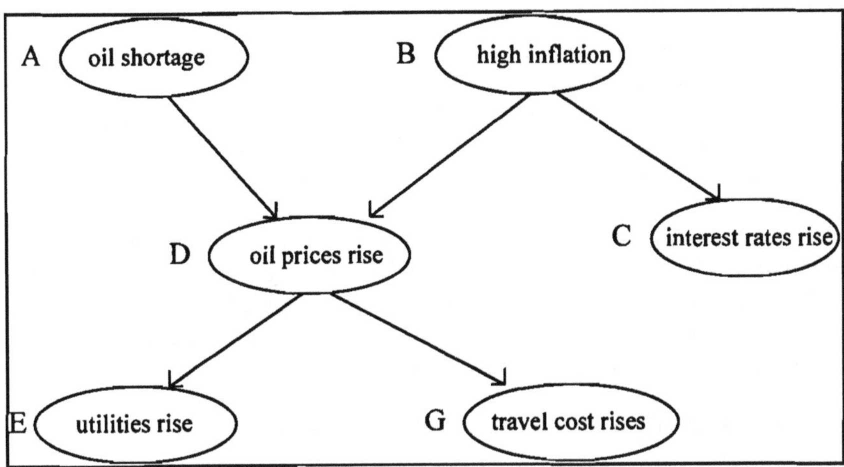

Figure 5.5 A Bayesian belief network.

When there is shortage of oil or high inflation then the oil prices rise. If the rate of inflation is high then the food prices rise. If the oil prices rise then the cost of utilities rises and the cost of travel rises.

The belief network corresponding to this knowledge is shown in Figure 5.5.

Normally, the probabilistic variables *A:"shortage of oil"* and *B:"high inflation"* are independent, P(B/A)=P(B), so there is no direct arc between them. But consider what happens if we know that oil prices rise and the inflation rate is high. This evidence should reduce the belief that there is a shortage of oil. So, B is dependent on *D:"oil prices rise"* and $P(B/A \wedge D) \neq P(B)$. In general, a graph search of the belief network can deduce such hidden dependencies and completely characterize all dependencies and independencies.

Consider the underlying undirected graph to the DAG (the DAG with edges without orientation). A node N in the undirected path from X to Y is called ***linear*** iff the path edges with N as an end point consist of one incoming and one outgoing edge from N.

The node N is ***converging*** if both edges are incoming to N and *diverging* if they both are outgoing from N.

A variable X is *dependent on Y given a set of variables W* iff there is an undirected path from X to Y with the following properties: Every node in the path which is not in W is linear or diverging or converging with a path ancestor in W. Every node on the path which is in W is converging.

For example, in the belief network of Figure 5.5, variable A is dependent on B given D and B is dependent on A given D. Variable E is not dependent on G given D, since in the path from E to G the node D is not converging. The joint probability of variables A,B,C,D,E,G is easily written down by inspecting the net topology:

$$P(A \wedge B \wedge C \wedge D \wedge E \wedge G) = P(D/A \wedge B)P(C/B)P(E/D)P(G/D)P(A)P(D)$$

In terms of computing the conditional probabilities of a variable given the values of a set of variables as evidence, it turns out that for belief networks with no loops, this can be done in polynomial time.

5.6.3 Propagating probabilities through a tree belief network

A *tree belief network* is a belief network where each node has at most one parent. In a *polytree belief network*, there is only one undirected path between any two nodes. In the case of tree or polytree belief networks, the probabilities can be propagated in polynomial time, eliminating the combinatorial explosion problem. In other types of networks, approximation techniques (for example using Monte Carlo analysis, [24] or Lauritzen's and Spiegelhalter's method, [16]) can be used for evaluating and propagating the probabilities.

For the case of a tree belief network, the propagation of probabilities proceeds by message passing between the nodes. Each node receives information from adjoining nodes to update its probability, and also passes information on to its parent and children. At the conclusion of the process, the net settles on a correct probability. The belief propagation process is described as follows (Pearl, [24]):

Consider the tree node X whose parent is U and children $Y_1,\ldots,Y_n$. Let e_X^+ be the evidence contained in the subtree rooted at U and $e_{Y_i}^-$ the evidence contained in the subtree rooted at Y_i. The parent of X sends as message to X the quantity $P(u/e_X^+)$ (the *causal support* to X) and the child Y_i of X sends to X as message the value $P(e_{Yi}^-/x)$ (*diagnostic support* to X). The total belief to X is updated using the formula:

$$bel(x)=P(e_{Y1}^-/x)...P(e_{Yn}^-/x)\sum_u P(x/u)P(u/e_X^+),$$

where u ranges over all values in the domain of variable U. The node X computes the following quantity, which it sends as a message to its parent U:

$$\sum_x P(e_{Y1}^-/x)...P(e_{Yn}^-/x)P(x/u).$$

Node X also computes the following quantity for each j and sends it to its child Y_j:

$$\sum_{u,x} P(x/u)P(u/e_X^+)P(e_{Y1}^-/x)...P(e_{Yj-1}^-/x)P(e_{Yj+1}^-/x)...P(e_{Yn}^-/x).$$

In case X is a root node, then it propagates to its children the prior probability of the root. If X is a leaf node, then it propagates to its parent the message (1,0) if X is true. If X is false, the message (0,1) is propagated upwards, and if X is not instantiated, then it propagates to its parent the message (1,1).

Propagating the evidence in the manner described above terminates in polynomial time in terms of the number of nodes in the net. For polynomial time algorithms on polytree networks and for approximation techniques to handle general network topologies, the reader is referred to Pearl, ([21]). For an implementation of Belief networks in Prolog, the user is referred to Shoham, ([28]).

One of the most widely known applications of belief networks is MUNIN, a system in the medical domain for interpretation of electromyographic readings, ([4]). The computations of posterior probabilities in MUNIN's belief network are extremely efficient even though the network is multiply connected and consists of around 1,000 nodes. For additional successful applications in the oil market, forecasting, ship classification and medicine, the reader is referred to [1], [15], [22], and [23].

In addition, a number of systems have been developed which implement uncertain inference using belief networks, for example the commercially available Hugin, ([3]), which implements Lauritzen's and

Spiegelhalter's method and Pulcinella which uses an algorithm by Shafer and Shenoy, ([27]). Pulcinella is a tool which can also dynamically build appropriate bayesian networks using available facts and relations expressed in predicate calculus format and resolution, ([25]). Finally, several algorithms have been developed for learning belief networks from raw data, see [14].

SUMMARY

This chapter described some of the most prominent approaches to dealing with uncertainty in expert systems. The probabilistic Bayesian approach seems to work well when applicable, but it has certain drawbacks. Each of the alternate models presented in this chapter were designed to address weaknesses of the Bayesian approach. Certainty factors are capable of better explanations of the reasoning path to a solution, fuzzy logic is better able to handle vagueness and imprecision, and the Dempster-Shafer theory can represent ignorance better. Finally, Bayesian belief networks eliminate many weaknesses of the pure Bayesian approach while maintaining many of its strengths.

EXERCISES

1. The truth value of a fuzzy implication is defined so that it extends the boolean implication. Show that out of the possible definitions for a fuzzy implication membership function in Section 5.5.3 only $\mu 1$, $\mu 2$, and $\mu 5$ behave exactly like the boolean implication, i.e.,
   $\mu_A(x)=0$ or $\mu_B(y)=1$ iff $\mu_{A \Rightarrow B}(x,y)=1$
   $\mu_A(x)=1$ or $\mu_B(y)=0$ iff $\mu_{A \Rightarrow B}(x,y)=0$

2. Write a Prolog procedure *run* of arity two, *run(Query, CFtotal)*, which takes as input a query and returns the answer to the query as well as the overall confidence factor to the truth value of the answer. (Procedure *run* basically has to convert the query to a list and then call the total function defined in Section 5.3.4).

3. Extend the meta-interpreter defined in Section 5.3.4 for implementing the certainty factors approach into the inference mechanism of Prolog, so that it can also process queries with uninstantiated variable arguments. All the possible answers to such queries are to be given with the overall confidence factor associated

to each of them.

4. Prove that $CF(h,e)+CF(\neg h,e)=0$

5. Given the rule base:
   if (cloudy) then (rain), CF=0.7
   if (warm) and (early summer) then (rain), CF= -0.9
   if (sunny) then (warm), CF=0.8

   and the facts:
   cloudy with CF=0.4
   sunny with CF=0.7
   early summer with CF=0.9

   Calculate CF(rain).

6. Let H={flu, cold, pneumonia} be the set of possible outcomes of diagnosis. Given the rules:
   IF fever THEN flu
   IF fever THEN pneumonia
   IF sneezing THEN cold
   IF sneezing then pneumonia
   a) Assume the confidence factors: CF=0.99 for the first rule, CF=0.05 for the second rule, CF=0.90 for the third rule, CF=0.01 for the fourth rule, and CF(fever)=1, CF(sneezing)=1. Apply the CF approach to determine which disease constitutes the better diagnosis.
   b) In the presence of the evidence "fever" the following mass assignment is given to the power set of the frame of discernment by the expert:
   $$m_1(\{flu\})=0.99, m_1(\{pneumonia\})=0.05, m_1(H)=0.05.$$
   In presence of the evidence "sneezing" the mass assignment is:
   $$m_2(\{cold\})=0.9, m_2(\{pneumonia\})=0.01, m_2(H)=0.09.$$
   Using the DS approach calculate the beliefs to each disease and the belief intervals.
   c) Compare the two results. Do you observe anything odd?

7. Consider as frame of discernment the set H={flu, cold, pneumonia}. Given the evidence "fever" an expert assigns these mass probabilities:
   $m_1(\{flu, pneumonia\})=0.8$, $m_1(\{cold\})=0.1$, $m_1(\{flu, cold\})=0.1$.
   Given as a second symptom "shivering" the expert may assign the mass probabilities:

m_2({pneumonia})=0.4, m_2({pneumonia, cold})=0.5, m_2(H)=0.1.
Use the combination rule to update the masses, compute the beliefs of the focal elements and the certainty intervals.

8. Define fuzzy sets for the terms "close_to(X,Y)", and "tall" where the domain is height in feet. Also construct membership functions for the fuzzy quantifiers "very" and "more or less" and the fuzzy truth values "true", "fairly true", "very true".
   a) Consider the rule:
      If (X is tall and his height is close to seven feet) then (he plays in the center position in basketball).
   Given that Christian is 6.5' feet, then should he play center?
   b) What is the truth value of the statements:
      It is fairly true that Christian is very tall and close to 7 feet.
      It is very true that Christian is more or less tall and close to 7 feet.

9. Consider the decision tree example given in Figure 5 of Appendix A. Introduce uncertainty by assigning a confidence factor to each rule corresponding to leaves in the tree and renaming the decision nodes as: high income, low income, big mortgage, small mortgage, young age, old age.
   a) Given the evidence that: Basil has high income with CF=0.99, Basil is young with CF=0.8, Basil is single, CF=1, Basil has a big mortgage with CF=0.2, derive the CF for each possible classification, high risk, moderate risk or stable investor.
   b) Use the Prolog procedures in Section 5.3.4 to write the knowledge in Prolog format and derive the CFs, by querying the Prolog program.
   c) Introduce fuzzy predicates high (or big) and low (or small) defined on the domain "amount of money", and young, old defined on the domain "years of age". Convert the decision tree to a set of fuzzy rules. Given the evidence: Basil's income is 400,000, he is 30 years old, owes 20,000 on his mortgage and he is single, use the fuzzy implications approach to determine his high, moderate or stable risk investor classification.
   d) Use the fuzzy logic Prolog meta-interpreter given in Section 5.5.4 to implement this fuzzy knowledge base in Prolog.

REFERENCES

[1] Abramson, B., "ARCO1: An Application of Belief Networks to the Oil Market", *Proceedings of 7th Confer. on Uncertainty in AI*, LA, 1991, pp. 1-8.

[2] Aikins, J.S., "Puff: An Expert System for Interpretation of Pulmonary Function Data", *Computers and Biomedical Research*, 16, 1983, 199-208.

[3] Andersen, S.K., Olesen, K.G., Jensen, F.V., Jensen, F., "Hugin: A shell for building belief universes for expert systems", *Proceedings 11th Int. Joint Confer. on AI*, pp. 1080-1085, Sydney, 1989.

[4] Andreassen, S., Woldbye, M., Falck, B., Anderson, S., "MUNIN: A causal probabilistic network for the interpretation of electromyographic findings", *Proceedings of 10th Intern. Joint Confer. on Artificial Intelligence*, pp. 366-372, 1987.

[5] Baldwin, J.F., Zhou, S.Q., "A Fuzzy Relational Inference Language", *Fuzzy Sets and Systems*, 14, pp. 155-174, 1984.

[6] Baldwin, J.F., "Evidential Support Logic Programming", *Fuzzy Sets and Systems*, 24, pp. 1-26, 1987.

[7] Bouchon-Meunier, B. "Inferences with inaccuracy and uncertainty in ESs" in: *Fuzzy Expert Systems*, Ed. A. Kandel, CRC Press, 1992, pp.43-54.

[8] Buchanan, B.G., Shortliffe, E.H., *Rule Based Expert Systems: The MYCIN experiments of the Stanford Heuristic Programming Project*, Addison Wesley, 1984.

[9] Buckley, J.J., Siler, W., Tucker, D., "FLOPS: A fuzzy expert system applications and perspectives", in: *Fuzzy Logics in Knowledge Engineering*, Negoita, C.V. and Prade, H., Eds., Verlag, 1986.

[10] Cohen, P.R., *Heuristic Reasoning about Uncertainty: An AI approach*, Pitman, Boston, 1985.

[11] Dempster, A.P., "Upper and Lower Probabilities Induced by

Multivalued Mappings", *Annals of Math. Stat.*, 38, 1967, 325-329.

[12] Fagan, L.M., "Extensions to Ruled based formalism for a monitoring task", in: *Rule Based Expert Systems*, Buchanan, B.G., Shortliffe, E.H., eds., Addison Wesley, 1984, 397-423.

[13] Gordon, J., Shortliffe, E., "The Dempster-Shafer Theory of Evidence", in: *Rule Based Expert Systems*, Buchanan, B., Shortliffe, E., Eds.,1985, pp. 272-292.

[14] Heckerman, D., "Bayesian Networks for Knowledge Representation and Learning", *Advances in Knowledge Discovery and Data Mining*, Fayyad, U.M. et al., eds, MIT Press, 1995.

[15] Heckerman, D., Mamdani, A., Wellman, M., "Real-World Applications of Bayesian Networks", *Comm. ACM*, vol. 38, 3, 1995.

[16] Lauritzen, S.L., Spiegelhalter, D.J., "Local Computations with Probabilities on Graphical Structures and their application to Expert Systems", *J. Royal Statistical Soc.*, 50, 2, 1988, pp. 157-224.

[17] Leung, K.S. and Lam, W. "Fuzzy concepts in expert systems", *Computer*, 43, September, 1988.

[18] Lowrance, J., Garvey,T., "Evidential Reasoning: A developing concept", *Proceedings of Intern. Confer. on Cybernetics and Society*, 1982, pp. 6-9.

[19] Mamdani, E.H., Gaines, B.R., *Fuzzy reasoning and its applications*, Academic Press, 1981.

[20] Martin, T.P., Baldwin, J.F., Pilsworth, B.W., "The implementation of FProlog-A Fuzzy Prolog interpreter", *Fuzzy Sets and Systems*, 23, 1987, pp. 119-129.

[21] Miyamoto, S., Yasunobu, S., "Predicate fuzzy control and its application to automatic train operation systems", *Proceedings Fuzzy Infer. Soc. Confer.*, 1984.

[22] Musman, S.A., Chong, L.W., "A study of scaling issues in Bayesian Belief Nets for ship classification", *Proceedings of 9th Confer. on*

Uncertainty in AI, 1993.

[23] Olesen, K.G., Kjaerulff, U., Jensen, F., Falck, B., Andraessen, S., Andersen, S.K., "A MUNIN network for the median nerve-A case study in loops", *Applied Artificial Intelligence*, 3, 1989, pp. 385-404.

[24] Pearl, J., *Probabilistic Reasoning in Intelligent Systems: Networks of Plausible Inference*, Morgan Kaufmann, San Mateo, CA, 1988.

[25] Saffiotti, A., Umkehrer, E., "Pulcinella: A general tool for Propagating uncertainty in valuation networks", *Proceedings 7th Confer. on Uncertainty in AI*, 1991, pp. 323-331.

[26] Shafer, G., *A Mathematical Theory of Evidence*, Princeton University Press, 1976.

[27] Shenoy, P.P., Shafer, G., "Axioms for probability and belief function propagation", *Uncertainty in AI*, 4, Eds. Shacter, R.D., Levitt, T.S., Kanal, L.N., Lemmer, J.F., North Holland, 1990.

[28] Shoham, Y., *Artificial Intelligence Techniques in Prolog*, Morgan Kaufmann, San Fransisco, 1994.

[29] Shortliffe, E.H., *Computer Based Medical Consultations: MYCIN*, American Elsevier, 1976.

[30] Van Le, T., "Fuzzy Programming in Prolog", *AI Expert*, 9, 7, 1994, p. 32.

[31] Yager, R.R., "Approximate reasoning as a basis for rule based Expert Systems", *IEEE Trans. Sys., Man, Cybern.*, 14, 6-36, 1984.

[32] Zadeh, L., "Review of books: A mathematical theory of evidence", *AI Magazine*, 81-83, Fall 1984.

[33] Zadeh, L., "Fuzzy Sets", *Information and Control*, 8, 1965, pp. 338-353.

[34] Zadeh, L., "The role of Fuzzy Logic in the management of uncertainty in Expert Systems", *Fuzzy Sets and Systems*, 11, 1983, pp. 199-228.

CHAPTER 6

MACHINE LEARNING

6.1 INTRODUCTION

Learning is a necessary ingredient of intelligence and it probably has to be one of the main characteristics of any intelligent computer system. Machine Learning is a field of study which has as its goal to enable computer systems to learn to create new knowledge, update existing knowledge and improve their performance without intervention and reprogramming.

Learning in an expert system can take two forms: automate the creation of the knowledge base or update and improve the knowledge base in view of new information. Learning in first generation expert systems was exercised in its most rudimentary form: learning by being told. This form of learning gave rise and magnified the knowledge acquisition bottleneck problem. As described in Chapter 4, the elicitation of knowledge from a domain expert is a slow, tedious and error prone process. The knowledge engineer had to painstakingly build and maintain the knowledge base. The expert system would then convert implicit knowledge contained in the knowledge base into explicit knowledge by using deductive reasoning techniques. Chapter 4 addressed a number of semi-automated techniques for assisting the knowledge engineer or the expert himself with the knowledge acquisition process.

Significant amount of work has been done in the area of machine learning and knowledge acquisition in automating the knowledge engineering process. Techniques derived from the area of machine learning may be able to more effectively address the knowledge acquisition bottleneck in expert system development. The ultimate goal of this research is to eliminate the need for knowledge engineers or at the very least provide the necessary tools to assist the knowledge engineering process.

A learning subsystem, which may augment a second generation expert system, consists of the *performer* module, the *learner* module, the *knowledge base* and the *critic*. The performer module encapsulates the inference engine. At each learning cycle, the current knowledge base is activated by the *performer*, usually after some external input. The *critic* evaluates the output of the performer on accuracy of deductions and performance. The *learner* uses the critic's evaluation to modify or extend the knowledge base to improve performance.

For a detailed introduction to the field of machine learning, the reader is referred to ([9, 14, 19, 27]).

6.1.1 Inductive, analogical and explanation based learning

The most prevalent machine learning paradigms include *inductive learning, analogical reasoning*, and *explanation based learning*.

First generation expert systems lacked the capability to learn or improve the knowledge base by themselves. Even though direct instruction, memorization and learning by being told can be viewed as less sophisticated learning methods, they are hardly adequate for addressing the knowledge acquisition bottleneck.

Analogical reasoning involves learning new concepts or deriving new solutions through use of similar concepts and their solutions. For example, Greiner's early NLAG system ([12]) accepts as input an incomplete domain theory and it conjectures solutions based on derived analogies. Other analogical reasoning systems include the ones based on derivational and transformational analogy (Carbonell, [3]).

A form of analogical reasoning, called *case-based reasoning*, has recently found enormous popularity among expert system developers and is further discussed in Section 6.5.

In *explanation based learning*, knowledge is derived from a single training instance by explaining why the given instance is an example of the concept to be learned. For an introduction to explanation based learning, see Mitchell et al., ([18]).

In *inductive learning*, concepts are learned from a number of positive and negative instances of the concept. The set of available instances is typically divided into a *learning set* (*training set*) and a *testing set*. The training and testing sets typically consist of ordered pairs {(x,c)}, where x is an instance vector in an attribute space X and c represents the index of the class that the instance belongs to. The attribute space X consists of the set of attributes which are used to describe the objects to be classified. Objects are described by their values for these attributes. The training set is the input to the learning system and is used to derive the generalized knowledge, which is then tested for accuracy and performance using the testing set of patterns.

The output of an inductive learning system is usually given as a *classification* or *decision rule.* A classification or decision rule, D(x), is a function which maps every instance x in X into some class c in the set C of class indexes. Generally, a classification rule maps an instance x into a probability distribution $(P_1, P_2,...,P_n)$, where P_i is the probability that vector x belongs to class i. Classification rules are often expressed in first order predicate calculus form. For example, a description for the concept "profitable_company" may be "a company with no competition, whose age is new and which produces software" or in Prolog form:

profit(X,up):-competition(X,no), age(X,new),product(X,software).

Inductive learning techniques include *decision tree* techniques, (Quinlan's ID3, [24], Breiman's et al. CART, [2], Cestnik's et al. ASSISTANT, [4]), *version space search,* (Mitchell's candidate elimination algorithm, [17], Clark and Niblett's CN2, [5]), *conceptual clustering techniques,* (Michalski's CLUSTER/2, [16]), the *evolutionary approach,* (Forsyth's BEAGLE, [8], Goldberg, [11], Holland, [13]), and the *connectionist approach.* Successful applications of inductive learning techniques, include tools for predicting sales in retail chain stores, for inventory management, predictions of customer response to direct-mail offers, learning rule bases for insurance underwriting and character recognition by automatic teller machines.

The basic example of a decision tree learning algorithm is Quinlan's ID3 algorithm ([24]) which is described in Section 6.2. ID3 chooses an attribute and subdivides the original set of training examples into subsets so that all patterns in a subset share the same value for that attribute. The information theoretic measure of entropy is used as a heuristic to determine which attributes to choose and in what order they should be used to subdivide the set of training examples. A subset of training examples is not subdivided any longer when all its elements share the same value of the goal attribute. The result of the algorithm is a decision tree, where all instances in a leaf node agree on their value of the goal attribute. The presence of errors in the data tends to have the effect of over-expanding the decision tree, producing long and unnecessarily detailed rules. Techniques for handling noisy domains are described in Section 6.3.

A system which is representative of the *version space search* and *conceptual clustering* techniques is described in Section 6.4. In this approach, the partially ordered set of domain concepts is searched for a most general concept describing all the positive patterns in the learning set.

The *evolutionary approach* is based on genetic algorithms and classifier systems and is discussed in Section 6.6. Genetic algorithms is a search and optimization technique which is based on the mechanics of natural evolution and genetics.

The *connectionist* (neural network) approach to machine learning is another major paradigm of learning from examples and it is explained in Chapter 7. A disadvantage of neural networks is that they are presently used as black boxes. Based on a training set of instance vectors, they can learn to classify patterns they have not seen before by adjusting their weights, but they lack the explanation facilities of rule based expert systems. As a result, a neural network cannot explain or justify the decisions it reaches. Currently, many research efforts are directed at providing neural networks with explanation facilities (Gallant, [10], Nikolopoulos, [23]). The main approach taken is that of using the network weights and topology to construct a rule knowledge base, which will perform at a comparable classification accuracy as the neural net. This interplay of neural networks and expert systems is discussed in Chapter 7.

Finally, Chapter 8 is concerned with how the symbolic, evolutionary and connectionist approaches are being merged into Hybrid expert systems in an attempt to render more versatile and powerful expert systems.

6.2 DECISION TREES AND THE ID3 ALGORITHM

The Iterative Dichotomizer 3, (ID3), algorithm is an inductive learning algorithm developed by Quinlan, ([24]), which deduces generalized knowledge from a set of training instances. The derived knowledge is expressed in the form of classification (decision) trees. Variations of ID3 have been successfully used for many applications including the learning of the knowledge base for systems which learn chess end games and predict soybean disease classifications. Many of the first commercial automated knowledge acquisition systems, like the VP-Expert, Xi Plus, 1st-Class, ExpertEase and Knowledge Quest were based on variations of the ID3 algorithm.

6.2.1 The ID3 algorithm

Given a goal attribute, (an attribute for which we want to learn how the other attributes affect its value), the ID3 algorithm constructs a classification tree in a recursive, top down, divide and conquer fashion, as follows:

Begin with the whole set of instance vectors at the root node and progressively separate them into subsets, one subset for each value in the domain of the chosen "branching" attribute, in such a way that all instances in the same subset share the same value of the "branching" attribute. Each generated subset forms a child node. If the instances contained in a node do not share the same value for the goal attribute, the node is expanded further by using another branching attribute, otherwise it becomes a leaf. The process terminates when there are no nodes that can be expanded further.

To keep the branching factor of the classification tree manageable, a continuous or multivalued discrete domain is usually divided into a small number of subdomains. For example, if the age of a company in the data set ranges between one and ninety years, we may want to convert it into three subranges, *old, average and new,* where "new" is any age less than ten years, "old" any age greater than 60 years and "average" ranges in between.

We give below an example of how a classification tree would be constructed for the following sample loan approval database.

Suppose we want to determine if an individual's loan application should be approved or not based on the determining attributes *income, home ownership* and *debt.* By constructing a knowledge base of rules with hypothesis the goal attribute "loan" and premises conditions on the other attributes, a "profile" of the successful applicant will be constructed.

Of course, statistical techniques could be used or a neural net could be trained on the data to predict the eligibility for a given applicant, but this approach will lack explanation facilities and will not provide any better understanding of the underlying relations between attributes and how they affect each other, nor does it allow the justification to the applicant of a recommendation to approve or reject the loan.

Consider the following training instances:

LOAN	INCOME	HOME-OWNER	DEBT
no	low	no	high
no	low	no	low
no	low	yes	high
no	average	yes	high
no	average	no	low
no	average	no	high
yes	average	yes	low
no	high	no	low
no	high	no	high
yes	high	yes	high

Let us now apply the ID3 classification tree process on the database given above. Suppose that we first choose the attribute *debt* as the branching attribute. Based on this attribute, we split the set of data tuples into two groups, one with "debt" equal to high and one with "debt" equal to low.

Both child nodes need now to be split, since there is no consensus on loan approval being yes or no in either leaf. If *income* is chosen next as the branching attribute, we get the second level of the tree shown in Figure 6.1. Four out of the six leaf nodes need not be expanded further, as the vectors in the subsets corresponding to each node share the same value for the goal attribute.

If "home-owner" is chosen to expand the remaining two leaf nodes, the classification tree of Figure 6.2 will be derived.

Each path from the root to a leaf in Figure 6.2 corresponds to a

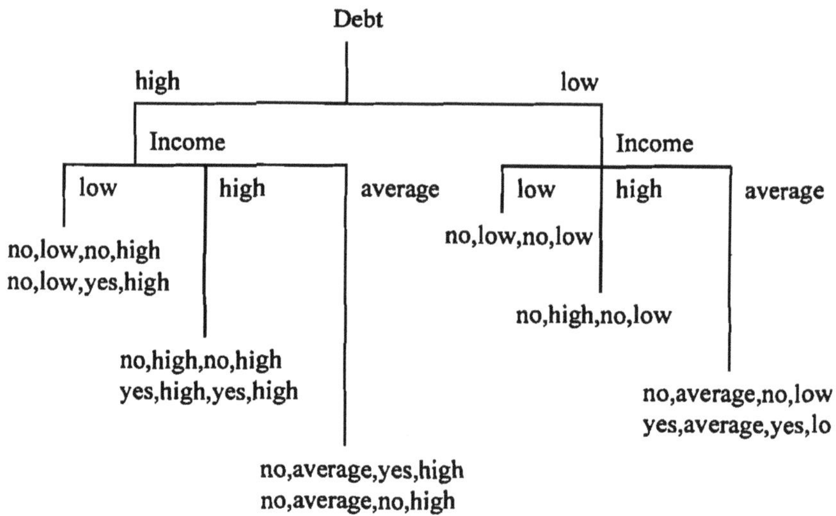

Figure 6.1 Expansion of the classification tree using Debt and Income.

rule. The rules derived from the tree given above are as follows:

loan(no):-debt(high),income(low).
loan(no):-debt(high),income(average).
loan(yes):-debt(high),income(high),home_owner(yes).
loan(no):-debt(high),income(high),home_owner(no).
loan(yes):-debt(low),income(average),home_owner(yes).
loan(no):-debt(low),income(average),home_owner(no).
loan(no):-debt(low),income(low).
loan(no):-debt(low),income(high).

It should be clear at this point that the choice of the branching attributes affects the resulting decision tree. For example, if we choose to first branch on "debt" followed by "home-owner" the decision tree of Figure 6.3 will be derived. This tree corresponds to the rule set:

loan(no):-debt(high),home_owner(yes),income(average).
loan(no):-debt(high),home_owner(yes),income(low).
loan(yes):-debt(high),home_owner(yes),income(high).
loan(no):-debt(high),home_owner(no).
loan(yes):-debt(low),home_owner(yes).
loan(no):-debt(low),home_owner(no).

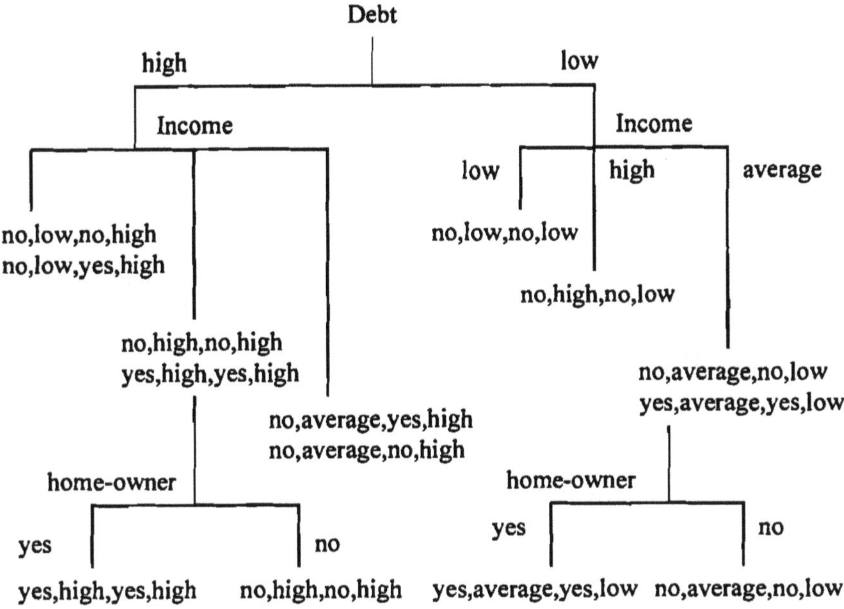

Figure 6.2 Further expansion of the classification tree using "home owner".

Both rule sets have a classification accuracy of 100 percent on the vectors in the data set, but in the second set of rules, the rules are shorter therefore more general, and there are fewer of them. The question now arises, how to determine the order in which to consider the branching attributes so that the derived tree is one of the smallest possible depth and branching factor. The brute force approach for finding the optimum tree, by producing the trees for all possible permutations of branching attributes and selecting the "best," causes combinatorial explosion. The problem of optimizing a decision tree is an NP complete problem. The ID3 algorithm proposes a heuristic solution to this problem. The information theoretic measure of entropy is used in order to select the branching attribute that results in the optimum tree.

The concept of entropy is introduced in the following section.

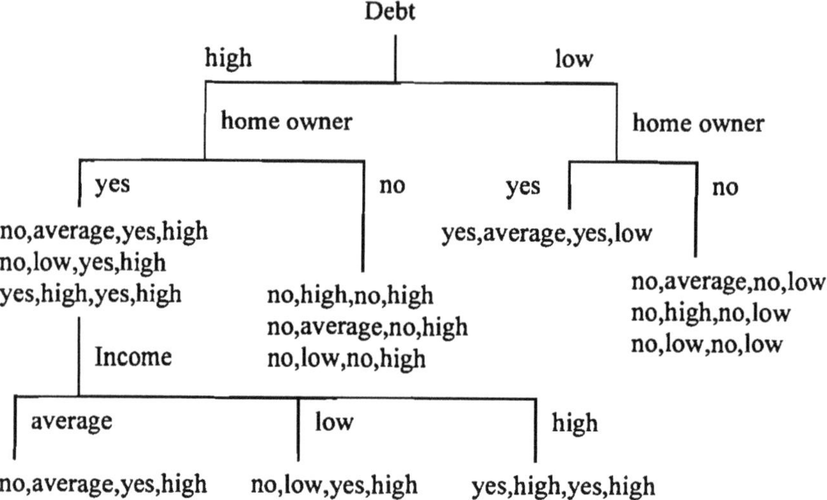

Figure 6.3 An alternate expansion of the classification tree.

6.2.2 Entropy

Given a completely expanded classification tree and a set of attribute values corresponding to a path from the root to a leaf, the "uncertainty" as to the value of the goal attribute has been reduced to zero. When no attribute values are known for an instance we have maximum uncertainty about its classification, while the more attribute values are known the more the uncertainty is reduced. In order to derive a classification tree of the smallest depth, a greedy heuristic can be employed where the attributes which cause the greatest reduction in uncertainty of classification are used first to expand the tree. ID3 uses the concept of *entropy* as a quantitative measure of the "uncertainty" of classification induced by an attribute.

The uncertainty, as to which of several events is going to occur (for example, whether the loan application will be approved or not), is a function of the probabilities of those events. In what follows, let us assume that $H(P_1,P_2,...,P_m)$ is such an uncertainty function, where P_i's are the probabilities of m events.

An uncertainty measure must satisfy the following requirements (Ash, [1]):

1. If the number of equiprobable choices for the outcome of an experiment is increased, then the uncertainty about the outcome is increased. For example, $H(1/2,1/2)<H(1/4,1/4,1/4,1/4)$

2. The uncertainty as to the joint occurrence of two independent events equals the sum of the uncertainties as to their individual occurrences.

3. The uncertainty measure must be a continuous function of the probabilities involved. In other words, a small change in the probabilities must be associated with a small change in uncertainty.

It has been proven that there is only one function which satisfies the given conditions, namely the entropy.

Given m possible outcomes of an experiment, having respective probabilities of $P_1,P_2,\ldots,P_m$, the uncertainty function is given by:

$$H(P_1,P_2,\ldots,P_m)= -\alpha(\sum_i P_i*\log(P_i))$$

where α is an arbitrary positive number, and the logarithm base is greater than one. This function is known in information theory as the *entropy of an information source*. In the following examples, we use the value of 1 for α and a logarithmic base equal to 2.

The entropy of an event measures our uncertainty about the possible outcome of the event. This measure can be extended to tell us how much information is gained regarding an event Y if we are told that a second event X has occurred. If X_i is the ith possible outcome for X and Y_j the jth possible outcome for Y, and $P(Y_j /X_i)$ is the corresponding conditional probability, then the *conditional uncertainty* of an event is defined as follows:

$$H(Y/X)= -\sum_{i,j}P(Y_j /X_i)*\log[P(Y_j /X_i)]$$

It is always the case that $H(Y/X)\leq H(Y)$. If X and Y are independent, then $H(Y)=H(Y/X)$. In general, the difference between the uncertainties, $H(Y)-H(Y/X)$, is a measure of the amount of information contained in X about Y or viewed another way, how much influence the outcome of event X has on event Y.

6.2.3 Using entropy to optimize the classification tree

Applying the concept of entropy to the decision tree optimization problem, we find that if an object can be classified into one of several different groups, the entropy is a measure of the uncertainty of the classification of that object. Mathematically, if an object can be classified into N classes, $C_1,...,C_n$ and the probability of the object being in class i is $P(C_i)$, then the entropy of classification C is

$$H(C) = -\sum_i P(C_i)*\log(P(C_i))$$

What is needed, in order to determine the attribute on which to further subdivide a node, is the entropy (uncertainty) of classification after an attribute is used. The decision tree will then be extended using the attribute that results in the smallest entropy of classification.

The first step in calculating the entropy of classification, H(C/A), for a particular attribute A, is tô split the data table into subtables in such a way that tuples in a subtable have the same value of the partitioning attribute.

The entropy of each subtable, $H(C/a_j)$, is calculated for each value a_j of the attribute A. For each possible outcome a_j of attribute A, $H(C/a_j)$ is given by the expression:

$$H(C/a_j) = -\sum_i P(c_i/a_j)\log[P(c_i/a_j) + n]$$

The function $P(c_i/a_j)$ is the probability that the class value is c_i when the splitting attribute value is a_j. The number n is an adjustment factor inserted so that the log is defined even when $P(c_i/a_j)=0$ and $\log[P(c_i/a_j)]$ is not defined. The overall entropy of classification, H(C/A), is then given by the weighted average of the entropy for each value a_j of the attribute. Mathematically this is expressed as

$$H(C|A) = \sum_{j=1}^{m} P(a_j)*H(C|a_j)$$

where m is the total number of possible values for the attribute A.

We now illustrate the use of entropy for constructing decision trees. Consider the loan approval database. If the attribute "debt" is used to subdivide the patterns into two sets, one for each domain value of "debt", we get:

LOAN	INCOME	HOME-OWNER	DEBT
no	low	no	high
no	low	yes	high
no	average	yes	high
no	average	no	high
no	high	no	high
yes	high	yes	high
yes	average	yes	low
no	high	no	low
no	low	no	low
no	average	no	low

We can now calculate the entropy of each subtable:

H(loan/debt=high)=
-[P(loan=yes/debt=high)log(P(loan=yes/debt=high))
+P(loan=no/debt=high)log(P(loan=no/debt=high))]
= -[(1/6)log(1/6)+(5/6)log(5/6)] = 0.196

H(loan/debt=low)=
-[P(loan=yes/debt=low)log(P(loan=yes/debt=low))
+P(loan=no/debt=low)log(P(loan=no/debt=low))]
= -[(1/4)log(1/4)+(3/4)log(3/4)]=0.244

In order to find the entropy of the entire table after the split, H(loan/debt), we must take the sum of the entropy for each of the values of the attribute and multiply it by the probability that the value will appear in the table.

H(loan/debt)=
P(debt=high)*H(loan/debt=high)+P(debt=low)*H(loan/debt=low)
=(6/10)*0.196 +(4/10)*0.244= 0.215 .

If we perform these same calculations for the other attributes in our example, we find that:

H(loan/income)=0.180, and
H(loan/home_ownership)=0.120.

Since *H(loan/home_ownership)* yields the smallest entropy and thus the least uncertainty, "home_ownership" is the best attribute to select for the initial split. Indeed by using this attribute we get right away the very short rule

IF (home_ownership=no) THEN (loan=no).

At each node, this decision process is repeated, in order to determine which attribute is to be chosen for further expansion. The reader can verify that "income" should be selected next, followed by "debt", and that the derived classification tree corresponds to the rules:

IF (home_ownership=no) THEN (loan=no).
IF (home_ownership=yes and income=low) THEN loan=no,
IF (home_ownership=yes and income=high) THEN loan=yes,
IF (home_ownership=yes and income=average and debt=low)
THEN loan=yes
IF (home_ownership=yes and income=average and debt=high)
THEN loan=no.

6.3 LEARNING FROM NOISY DATA

In many applications, the learning data set is imperfect. One common problem is that errors in attribute or class values and missing values can occur. In such cases, we say that the data is noisy. Inducing decision trees from noisy data, using the ID3 tree induction algorithm, gives rise to two problems: first, induced trees may unreliably classify new objects, and second, induced trees tend to be large and thus hard to understand. Noise makes the learning task more difficult and in order to effectively learn in a noisy domain, the requirement that learned concept descriptions match all the positive examples and no negative ones, is usually relaxed. The learned description is permitted to misclassify some of the learning objects. As an example, consider a situation in which we are to extend a node of a decision tree and let S be the current subset of objects in the node. Suppose that there are 100 objects in S, 99 of which belong to class C_1 and one of which belongs to class C_2. Knowing that there is noise in the learning data set, and that all of the objects in S agree on the values of the attributes already selected up to this point, it seems plausible that the object in class C_2, is in S due to an error in the data. If so, it is best to ignore this object, terminate expansion of that node in the decision tree and make it a leaf labeled with class C_1. Since the original ID3 algorithm would in this situation further expand the decision tree, by stopping at this point we have pruned a subtree of the complete decision tree. Tree pruning is the key to coping with noise in tree induction algorithms. An algorithm may effectively prune decision trees by using some criterion that indicates whether to stop expanding the tree or not. The stopping criterion would typically take into account the number of examples in the node, the prevalence of the majority class at the node, etc. This kind of pruning, accomplished through stopping the tree expansion if certain criteria are met, is called *construction-time pruning* (*pre-pruning*). In an alternate method of pruning, called *post-pruning*, the whole decision tree is constructed first and then pruned, starting at the leaves and moving upwards.

6.3.1 Construction time pruning

Assume that the generated decision tree has become very "deep" (i.e., there are some long paths from the root to the leaves) and that most nodes, far away from the root, contain instances with the vast majority sharing the same value of the goal attribute, except maybe a few. This phenomenon indicates that those nodes should contain instances of a single class, and the small percentage of instances in the other class is probably due to noise. These nodes are created in the decision tree because the induction algorithm tries to overfit the training set. A good strategy is to stop branching when this situation occurs.

As an example of a criterion, which can be used to determine when a node should be pruned, consider the *depth pruning ratio*. The depth pruning ratio is defined as follows:

$$R(a,d)=a^d$$

where the depth pruning threshold, a, is a constant ranging between 0 and 1, and d is the depth of the node under consideration for expansion (i.e., the number of branches from the tree root to the current node). During the induction of a decision tree, the percentages of positive instances and negative instances are compared with the depth pruning ratio at each node. When either of them is greater than or equal to the depth pruning ratio, this node is labeled as a leaf without further branching.

Based on the definition of the pruning ratio, it may be necessary to keep partitioning attribute nodes having nearly 100 percent instances of one class, when the nodes are close to the root. This can prevent excessive pruning and therefore ensure accuracy, which is guaranteed by the small depth of these nodes. On the other hand, if those attribute nodes are far away from the root, branches below them are probably unnecessary. The depth pruning ratios decrease as the depth of the node increases. Eventually, the percentage of positive (or negative) instances will exceed the depth pruning ratios, and these nodes will be labeled as leaves.

It is crucial to have a good depth pruning threshold. A bad threshold will lead to either over-pessimistic pruning (e.g., a=1) or excessive pruning (e.g.,a=0). Whereas a good one can both prune the noisy branches effectively and avoid excessive pruning. Experimentation with different threshold choices is needed, until a satisfactory decision tree is derived.

For examples of other existing pruning techniques like retrospective pruning and the chi-squared method, the reader is referred to [4] or [5].

6.3.2 Post pruning

We present here a post pruning algorithm, which was introduced in the Assistant 86 learning system, ([4]). The key decision in post-pruning is whether to prune a subtree or not. This decision is based on classification error estimates of the decision subtree.

Let S be the set of objects classified into the root of the subtree considered for pruning, and K be the number of classes (size of the domain of the goal attribute). Suppose that there are N instances in S, the majority class in S is C, n out of N instances in S belong to C, and the rest of the instances in S belong to other classes. If we prune the subtrees of this node and label the node with its majority class C, the result will be the introduction of a classification error. Under the assumption that all classes are equiprobable and that the prior distribution of the probabilities of the classes is uniform, the expected probability of the classification error at that node E(S) for set S is:

$$E(S)= (N-n+K-1)/(N+K)$$

Such an error estimator can be directly applied to estimate the classification error at each leaf of a decision tree: Error(leaf)=E(S), where S is the set of instances at the leaf. This error estimator is easily generalized to estimate classification errors of non-leaf nodes in a decision tree as follows.

Let *Node* be a non-leaf node of a decision tree with children $Node_1$, $Node_2$, etc. Let the probabilities of branches from *Node* to $Node_i$ be P_i. The probabilities P_i can be approximated by relative frequencies of the sets of instances contained into nodes $Node_i$. Then we define:

$$\text{BackedUpError(Node)} = \sum_i P_i * \text{Error}(Node_i)$$

We can now compare the node's "static" error with its backed-up error. In the case that the static error is lower, then we prune the node's subtree.

This formula suggests the following pruning algorithm to optimize the estimated classification error. Given a complete (unpruned) decision tree, the algorithm starts at the bottom of the tree and progresses toward the root of the tree, backing up the error estimates and pruning inferior subtrees. This algorithm guarantees that all of the subtrees are pruned optimally with respect to the classification error estimates. As a result, it will minimize the expected error at the root of the tree; that is, the expected error of the whole tree. Of course, the pruned tree obtained this way is only optimal with respect to error estimates.

Figure 6.4 shows the pruning of a decision tree using the procedure described above. In this figure, the left-most leaf of the unpruned tree has class frequency [3,2], meaning that there are three objects of class C_1 and two objects of class C_2 contained into this leaf. The static error for these class frequencies is:

$$E = (N-n+K-1)/(N+K) = (5-3+2-1)/(5+2) = 0.429$$

Similarly, for the left-hand child of node b, the error estimate is 0.4 and for the right child the error is 0.2. For node b, the static error estimate is: E(b) = (11-9+2-1)/(11+2)=0.230.
The backed-up error estimate for b is:
backup_error(b)=(3/11)(0.4)+(8/11)(0.2)=0.254.
Since the static error of node b is smaller than the backup, the subtrees of node b will be pruned making b a leaf. The process is repeated for node a.

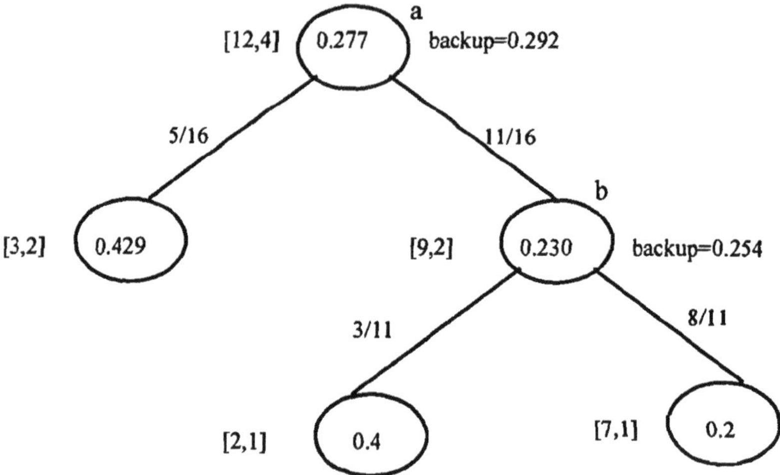

Figure 6.4 Postpruning an ID3 classification tree.

6.4 VERSION SPACE SEARCH AND CONCEPTUAL CLUSTERING

We present an algorithm, ([21]), which is a combination of the candidate elimination algorithm, ([17]), and the conceptual clustering technique ([16]).

Assume that each training example is expressed as a vector of attribute-value pairs. The attribute values can be nominal or categorical in nature.

An *object* is defined as an instance vector of attribute-value pairs. For example, O=(age=new, competition=no, product=tv, profit=up) is an object.

A *concept* is a generalization of an object and is described by the same description language as objects, but attributes can assume a set of values instead of just a single value. For example, C=(age={new,old}, competition=no, product={tv,vcr}, profit=up) is a concept, which contains the object O as an instance. Each concept corresponds to a rule. For example, given that the goal attribute is the profit, concept C can be stated as: If a company is old or new, and there is no competition, and the product is TVs or VCRs, then the profit is up.

A concept A is *more general than* a concept B or A *covers* B (A>B) iff each object which belongs to the concept B also belongs to the concept A. The relation "more general than" defines a partial order on the set of all possible concepts defined on an attribute set. The search techniques of AI can be used to search the lattice of concepts (the *version space*) in order to find a *most general*, *complete* and *consistent* concept covering the set of training examples.

A concept A is *complete* with respect to a set of training examples iff it covers all possible examples in the set.

A concept A is *consistent* with respect to a set of training examples iff it contains none of the negative examples.

A concept G is a *maximally specific generalization* of a concept A iff G is more general than A and there is no concept G' such that G>G'>A. Consider the concept A=(age=midlife, competition=no, product_type=VCR). Maximally specific generalizations can be obtained by assigning to one of the attributes one more value than the ones it already has. For example, the concept G=(age={midlife,new}, competition=no, product_type=VCR) is one of the maximally specific generalizations of A, since there is no concept less general than G which covers A.

The algorithm given below is a variation of the candidate elimination algorithm ([17], [19], [20]). It is a bottom up algorithm, in contrast to ID3. ID3 starts with the whole set of examples or equivalently with the most general concept, ( ), and proceeds by specializing. The algorithm given below, starts with positive instances (the most specific concepts) and gradually generalizes them, so that they remain consistent. Heuristics are used to guide the search through the potentially prohibitive size of the version space.

<u>*Algorithm*</u>:

Initialize the concept set (output) to empty.
REPEAT
    Get a positive example.
    IF the example is covered by any of the concepts in the current
        concept set THEN go to LOOP
    else
    begin

the example becomes the maximally specific concept at the root level of a tree.
Current_level=1
REPEAT
For each concept node in the Current_level find all its maximally specific generalization concepts, which become its children in the tree, at level:
Current_level=Current_level+1.
Delete the concepts from Current_level that are not consistent.
UNTIL all nodes are leaves
Insert all concepts corresponding to tree leaves in the concept set.
end
LOOP UNTIL no more positive examples.

If m is the number of attributes, then the size of a concept tree can be as big as $1+m+m(m-1)+m(m-1)(m-2)+...+m(m-1)...2.1 \approx m!$. To avoid the combinatorial explosion problem, heuristics can be used to guide the expansion of the trees. A fitness measure can be used to evaluate the potential usefulness of a node and only the best K nodes of a level are expanded each time. Let A be the attribute which will give the smallest uncertainty (entropy) of classification about the value of the goal attribute. The tree nodes, that will be expanded by the algorithm, will be the ones derived from their parent by extension of the attribute A.

Consider for example, the set of object vectors (from [26]), where the goal attribute is *profit=up*:

PROFIT	AGE	COMPETITION	PRODUCT TYPE
down	old	no	software
down	midlife	yes	software
up	midlife	no	hardware
down	old	no	hardware
up	new	no	hardware
up	new	no	software
up	midlife	no	software
up	new	yes	software
down	midlife	yes	hardware
down	old	yes	software

The algorithm would expand the first positive instance as follows:

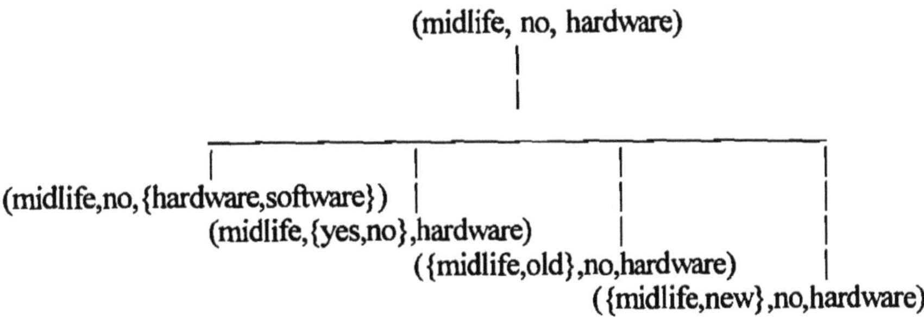

The second, third and fourth concepts at level 2 of the tree cover the negative instance (down,midlife,yes,hardware), so they are eliminated from further consideration. The first node is expanded by its maximally specific concepts as follows:

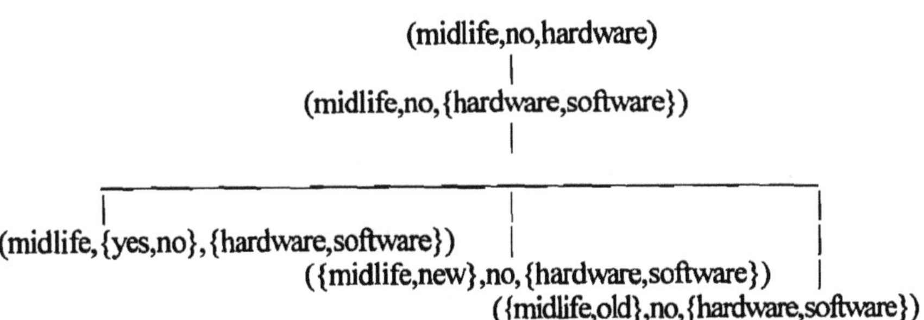

The first and third of the leaves are inconsistent (they cover negative instances), so the concept set after the first positive example is the concept ({midlife,new},no, {hardware,software}). This concept corresponds to the rule: *IF age=(midlife or new) and competition=no THEN profit=up*.

Repeating the process for each positive example, will also yield the rule: *IF age=new THEN profit=up*.

As another example, consider the following set of object vectors, where the type of airplane is the goal attribute.

PLANE TYPE	ENGINE TYPE	WING PLACED	WING SHAPE	TAIL	BULGES
C130	prop	high	regular	regular	under wings
C141	jet	high	swept	T-tail	aft_wings
C5A	jet	high	swept	T-tail	none
C747	jet	low	swept	regular	aft_cockpit

With a branching factor of K=1, and the entropy used as a heuristic measure, the algorithm will derive the same rule set as ID3, namely:

IF bulges=under_wings THEN airplane=C130
IF bulges=aft_wings THEN airplane=C141
IF bulges=none THEN airplane=C5A
IF bulges=aft_cockpit THEN airplane=C747

For K=2, three more rules will be derived in addition to the rules above:

IF wing_shape=regular THEN airplane=C130
IF engine=prop THEN airplane=C130
IF wing_position=low THEN airplane=C747

Finally we give an example where ID3 fails, but the conceptual clustering algorithm succeeds. Consider the following database containing contradictions:

DISEASE	EYES	TEMPERATURE	NOSE
none	clear	normal	clear
cold	bloodshot	high	runny
flu	bloodshot	high	runny

ID3 fails, while the algorithm described above gives the following set of rules:

IF nose=clear THEN disease=none (for branching factor K=1)

or

IF nose=clear THEN disease=none and
IF eyes=clear THEN disease=none (for branching factor K=2).

6.5 CASE BASED REASONING

Case-based reasoning (CBR) is a form of analogical reasoning which has enjoyed increased popularity as a developmental methodology due to its avoidance of the knowledge acquisition bottleneck and the natural way in which it integrates with object oriented environments. In *case-based reasoning*, the system recalls previous situations similar to the current one and adapts them to help solve the current problem. The existing descriptions of previously encountered problems are known as *cases*. Cases are usually implemented as objects in an object oriented environment even though other knowledge representation schemes (for example attribute-value pairs, rules or conceptual graphs) have also been used depending on the domain characteristics.

The development of a CBR system involves the collection of a depository of cases, deciding which features are important, and choosing an appropriate case representation scheme. The CBR system consists of the *case database*, and subsystems for *case indexing*, *case retrieval*, and *case adaptation*.

When a new case is presented, the case retrieval subsystem searches the indexed database of cases for the most similar case. The *adaptation subsystem* uses the retrieved case as a model in order to create a solution for the new problem, to warn the user of possible failures that have occurred in the past, and to interpret the current situation. Two types of adaptation are possible. In *structural adaptation* the solution of the retrieved case is modified to fit the new case, while in *derivational adaptation* a new solution is created by using the same process as used in the retrieved case. If no case can be retrieved which is sufficiently similar to the current case, then the current case together with a solution supplied by a human expert is stored and indexed by the case indexing subsystem into the database of cases, thus enabling the system to learn.

Several techniques are available for indexing a database of cases including *conceptual clustering, inductive indexing*, and *nearest neighbor indexing*. In *inductive indexing* an algorithm such as ID3 or Michalski's conceptual clustering is used to create a classification tree or a clustering which groups sets of similar cases together. When a new case is presented, the values of its attributes determine the path down the classification tree that should be followed or the cluster that contains the most similar cases to the new one. In the *nearest neighbor indexing* scheme, each feature of a case is assigned *match* and *mismatch weights* by the developer. The match weights indicate how important and necessary is the presence of this feature in order to call two cases similar. The mismatch weight indicates how important to similarity is the absence of this feature from one of the compared cases. The weights are numerically manipulated and used by various *similarity metrics* to retrieve the most similar case.

Several case-based expert system shells or shells with case-based subsystems have become available including KwEshell, ART-IM, CBR Express, Esteem, ReMind (see Appendix C). Application areas ideally suitable for CBR system development include the legal domain, intelligent tutoring, and systems design. For a thorough introduction to Case-based reasoning, the reader is referred to the book by Kolodner, ([14]).

6.6 EVOLUTIONARY MACHINE LEARNING

The evolutionary approach to machine learning is based on the concepts of genetic algorithms and classifier systems.

Genetic algorithms and classifier systems are an optimization, search and machine learning technique, suitable for a variety of problems, including NP-complete, and state-space type of problems. For a detailed introduction to the topic, the reader is referred to Davis, ([6]), Goldberg, ([11]) or Holland, ([13]).

6.6.1 Genetic algorithms

Genetic Algorithms are based on the concepts of natural selection and genetics. They were introduced by Holland, ([13]), with the intention of designing a computer system with many of the characteristics of natural systems.

A genetic algorithm is a global search technique as it does not look at a single point from where to expand the search for optima, but it evolves a population of points towards the optimum solution. This feature makes them particularly suitable for parallel processing and also reduces the probability of getting stuck at local optima.

A genetic algorithm solves a problem as follows. An initial set of possible solutions is chosen at random and the solutions are encoded as finite strings of symbols. The encoded set of possible solutions is called a *population*. The genetic algorithm does not need extensive knowledge of the problem domain in order to perform an efficient search of the solution space. It only requires a function (the *fitness function*) to measure the fitness of a candidate solution. The closer an encoded string is to being a solution, the bigger its fitness value should be. The initial population is successively transformed to populations of greater and greater overall fitness until a convergence criterion is met. During evolution the population size usually remains constant. The basic genetic operators used to transform a population to a better one are: *reproduction, crossover and mutation.*

In *reproduction*, the individuals in the population are propagated to the next population with a probability proportional to their fitness, in such a way that individuals of greater fitness have a bigger chance of surviving in the next population. Many alternate ways have been proposed to achieved this, with the simplest but still quite effective being the original one proposed by Holland. In this approach, usually referred to as the *roulette wheel* approach, the overall fitness of the population is calculated as the sum of the fitness of the individuals. The ratio of an individual's fitness to the overall fitness corresponds to a proportional slice of a roulette wheel. Which individuals make it to the next generation is determined by successive spins of the roulette wheel. For example, suppose the present population consists of the individuals 11000, 00111 and 10101 with corresponding fitness 30, 60 and 10. Then the spinning wheel is divided into three subareas, subarea 1, 2 and 3, corresponding to 30%, 60% and 10% of its total area. The wheel is spun three times to determine the three individuals in the next generation. If the wheel lands in area 1 then the bit string 11000 is reproduced, if it lands in area 2 then the string 00111 is reproduced, and in area 3 then 10101 is reproduced. Clearly the string 00111 is twice as likely as string 11000 and six times as likely as string 10101 to reproduce.

Crossover introduces variety in the population of candidate solutions by combining two existing individuals (*parents*) into new individuals (*offsprings*). Many forms of crossover are possible, and in its simplest form crossover works as follows. If k is the length of the two strings, randomly select a position p between 1 and k. The new strings are produced by merging positions 1 to p of parent 1 with positions p+1 to k of parent 2, and positions 1 to p of the parent 2 with positions p+1 to k of parent 1. For example, consider the individuals 110101 and 010011. If crossover is applied at position 3, the resulting offspring is 110011 and 010101.

Mutation introduces variation to the population by randomly altering individuals. This can lead the algorithm out of stagnation at local minima. One way to implement mutation is to randomly pick an individual and a position p between 1 and k, and change the value at position p to another domain value chosen at random.

Genetic algorithms can be analyzed by looking at how schemata behave during the evolution of the population from generation to generation. A *schema* is a set of encoded solutions. For example, the schema 11** represents all strings of length four, with the first two bits equal to 1. The *Fundamental Theorem of Genetic Algorithms* shows that short, highly fit schemata propagate themselves through evolution in a manner of exponential growth and converge to a solution ([11]).

Solving a problem using genetic algorithms, involves finding a way to encode the candidate solutions as finite strings, finding an appropriate fitness function to evaluate the goodness of a candidate solution and creating appropriate reproduction, crossover and mutation operators for the given problem. This process is illustrated by the following example.

Consider the boolean satisfiability problem (SAT). Given a boolean expression of n variables, is there an interpretation (assignment to the variables) that will make the formula true? This is an NP-complete problem (it cannot be solved in polynomial time. As the number of variables increases, the complexity increases exponentially). There is an obvious way to encode the candidate solutions as strings. In fact they are binary strings of fixed length. It takes more thought to find an appropriate fitness function. The obvious choice of assigning a fitness of 1 to a string that satisfies the boolean expression and 0 to one which does not, will not work. The majority of the population elements will have fitness equal to 0. Many variations of fitness functions for the SAT have been proposed in the literature, with one of them being the following. To evaluate the fitness of a binary string as a candidate interpretation to a boolean function, use the following formulas:

fitness(x)=1, for bit 1 and 0 for bit 0
$\text{fitness}(\neg x)=1-\text{fitness}(x)$
$\text{fitness}(x_1 \vee \ldots \vee x_n)=\max(\text{fitness}(x_1),\ldots,\text{fitness}(x_n))$
$\text{fitness}(x_1 \wedge \ldots \wedge x_n)=(1/n)\sum_i \text{fitness}(x_i)$

For example, consider the function:

$b(x_1,x_2,x_3,x_4)=(x_1 \vee x_2)\wedge(x_3 \vee \neg x_4)\wedge \neg x_1.$

In order to find a substitution which satisfies the formula, the genetic algorithm proceeds as follows:

Randomly pick a population of binary strings of length 4, say: 1011, 1001, 0010, 1101. For each string evaluate its fitness:

$\text{fitness}(1011)=(1/3)[\max(\text{fitness}(x_1),\text{fitness}(x_2))+\max(\text{fitness}(x_3), 1-\text{fitness}(x_4)) +(1-\text{fitness}(x_1))]$
$=(1/3)[\max(1,0)+\max(1,1-1)+(1-1)]$
$=(1+1+0)/3=2/3$

Similarly, fitness(1001)=1/3 and fitness(0010)=2/3.

The roulette wheel will be separated into three subsections of area 40%, 20% and 40% of the total area. During reproduction, the first and third strings will have double the chance to reproduce than the second string. Strings will randomly be chosen to be crossed over at a random position. For example, if string 1011 and 0010 are crossed over at position 2, they will produce the offspring 1010 and 0011. Mutation here is obviously necessary, as any eventual solution must have the second bit equal to 1, and this cannot happen from the initial population just by reproduction and crossover. As generation after generation of candidate solutions are produced, somewhere down the line, the random choice of which bit to mutate will be for position 2, introducing in the population the desired 1 in position 2. Parameters such as the population size and mutation rate (how often to apply mutation) affect the speed of convergence of the GA.

6.6.2 Classifier systems

Classifier systems have certain similarities with production systems as they are also based on match-select-act cycles. But unlike production systems, which typically select to fire one rule at a time, classifier systems can activate multiple rules in parallel. They also differ from production systems in that they employ competitive arbitration strategies to select which rules are to be fired and a reward based system for evaluating the strength of individual rules. The rule base of a classifier system continually evolves, through the use of genetic algorithms which evolve weak rules to rules of greater strength.

A classifier system is a machine learning system which consists of four main modules: a *classifier pool*, *a message list*, an *apportionment of credit algorithm*, and a *Genetic Algorithm*, ([11]).

The *classifier pool* is a set of production rules (*classifiers*) written in the form:

condition:action, strength

which represents the rule:

IF (condition) THEN (action): strength.

A single condition or a conjunction of conditions can be present in the classifier. The strength assigned to a rule is indicative of its usefulness to the system. The condition and action part of the rule are usually encoded as strings of a finite symbol set, for example {1,0,#}. Both conditions and actions are encoded as strings of the same length k. The # symbol corresponds to the "don't care" element. If it is present in a condition, then it can be matched by either a 1 or a 0 in the corresponding position of a message.

Messages (inputs) posted by the environment are written on a blackboard (message list). Classifiers, whose condition part matches a message on the blackboard, compete and the winners activate and post their action as a message on the blackboard or as output of the system. The process is repeated, producing a chain of classifier firings. When a classifier fires and it has the # symbol present in its action part, the symbol, in the corresponding position of the message that matched its condition, is assigned to the # symbol. For example, if the message list contains the message 1101011, the classifier 11#10#1:000#1## can fire and post the message 0001111.

An *apportionment of credit algorithm* is used to decide which classifiers will fire and also update the strength of classifiers in proportion to their usefulness. An example of such a system is the *Bucket Brigade Algorithm.* Initially, on the absence of any information about the usefulness of the classifiers, all of the classifiers are assigned equal strength. Classifiers, which are eligible to fire by matching a posted message to their conditions, bid an amount proportional to their strength for the right to fire. The winner classifiers (a set of classifiers posting the highest bids) fire and post their actions as messages on the blackboard. The strength of a firing classifier is reduced by the bid amount. The bid amount is divided, among the classifiers which posted the messages enabling the winner to fire, and is added to their strengths. The following difference equation determines how the strength of a classifier is updated in time t+1 (after a firing cycle) in terms of its strength in time t:

$$S(t+1)= S(t)-B*S(t)-T*S(t)+R(t)$$

where B is the *bid rate* with range $0 \leq B \leq 1.0$. If the classifier did not fire, then B=0. R(t) is the reward the classifier is due for posting any messages enabling others to fire and T is the *tax rate*, $0 \leq T \leq 1$. Each classifier pays to the environment an amount proportional to its strength, to avoid predominance of the rule population by the stronger rules. Usually T is chosen much smaller than B, and in such a way that $0 \leq T+B \leq 1$. It has been shown that the difference equation given above is stable, in the sense that as t approaches infinity S(t) approaches 0.

The quality of the rules in the classifier pool is improved by applying *genetic algorithms* to the population of classifiers, with the classifier strength used as their fitness value. The genetic algorithm can be called to improve the classifier pool, either probabilistically, with an average number of time steps between calls, or deterministically, with a fixed number of time steps between calls.

6.6.3 Deriving or improving rule sets with genetic algorithms, BEAGLE

One of the first systems, to use genetic algorithms for deriving rules from data, was the BEAGLE system (Biologic Evolutionary Algorithm Generating Logical Expressions, Forsyth, [8]). Given a set of training examples, of the form ($attribute_1$=$value_1$,..., $attribute_n$=$value_n$), where $attribute_n$ is the goal attribute, we wish to derive rules specifying under what conditions on the other attributes, the goal attribute will assume a given value.

For example, consider the company data of Section 6.2.1. Assume that the possible domain values for the attributes are:

for company age: 1 (very old), 2 (old), 3 (midlife), 4 (new), 5 (very new)
for product type: 1 (software), 2 (hardware).
for competition: 1 (if there is competition), 0 (if there is no competition).
for profit: 1(if profit is up), 2 (if profit is down).

The objective is to derive rules of the form:

IF conditions THEN profit=1.

The Beagle algorithm randomly picks a population of rule conditions. For example:

condition 1: (age<=3)
condition 2: (product type<2)∧(competition=1).

The first condition corresponds to the rule:

IF company is very old or old or midlife, then the profit is up.

The second condition corresponds to the rule:

IF product is software and there is competition, then the profit is up.

Possible positions for crossover are the relational operators (<, >, etc.) and logical operators (¬, ∧, ∨). In the conditions above, if the positions of <= and < are randomly chosen for a crossover, the offspring conditions after crossover will be *(age<=2)∧(competition=1)* and *(product type<3)*. These conditions correspond to the rules:

IF company is old or very old and it faces competition then profit is up
and
IF product is software or hardware THEN profit is up.

If, on the other hand, the relational operator <= and the logical operator "∧" were picked as the crossover points, then the following two conditions would be produced after crossover:

age<=(competition=1)
and
product type<2 and 3 which reduces to: *product type<2.*

Semantically, these conditions correspond to the rules:

If the age is very old and competition is present then profit is up
and
If product type is software then profit is up.

The genetic algorithm of Beagle produces the next population of rules from the present population as follows.

Each rule is evaluated on all the training examples and its fitness calculated in terms of the number of examples correctly classified by the rule. The rules are ranked in descending order of fitness and the bottom half of weakest rules is removed. Pairs of surviving rules are randomly crossed over, as described above, and the offsprings inserted in the population. Finally, a few rules may be mutated at random by changing the value of an attribute or its value in the rule. The process is repeated, producing generations of rules, until a determined number of iterations is reached or rules of satisfactory classification accuracy are reached.

Beagle has successfully been used to produce the knowledge base for a number of different applications, including a system to classify glass fragment evidence in forensic science (Evett and Spiehler, [7]).

6.6.4 Deriving a knowledge base for insurance underwriting using genetic algorithms

Below, we give another example of how genetic algorithms were used to automatically derive a knowledge base for insurance underwriting (Nikolopoulos, Duvendack, [22]). The machine learning component of the described expert system extracts underwriter rules from a collection of policy holders' records, in order to determine when a policy should be cancelled.

Policy holders' records are given in the form of attribute-value

pairs. One of the goal attributes is the "*terminate_policy*" attribute which assumes one of the values:
terminated_by_company, *terminated_by_policy_holder* and *active_policy*. Each data record had 19 attributes. The training set was noisy, mainly because of missing attribute values. The data set was separated into a training and a testing set, containing 70% and 30% respectively of the data patterns available.

Each individual of the genetic population consists of a sequence of rules and represents a candidate knowledge base for the problem domain. The 19 attributes were labeled from 1 to 19 and each rule in an individual is of the form: $a_1,a_2,...,a_{19},g_i$ where a_i belongs to the domain of attribute i or it is equal to *, and g_i is a value of the goal attribute. The value of * corresponds to the don't care symbol, and was introduced in order to maintain the individual's length as a constant, while at the same time permitting rules of varying lengths. The presence of * in position i of the rule chromosome indicates that attribute i is not present as a premise in the rule. For example, the encoding 2*0**...*3 corresponds to the rule:

IF (number_of_claims=2 and underwriter_code=no-violations) THEN policy=active,

since number_of_claims was the first attribute in the attribute labeling, and underwriter_code was the third. The value 3 was used to encode the value "active" for the goal attribute policy.

The given encoding enables all rules to be of constant length. The number of rules in an individual is also held constant and needs to be supplied to the algorithm. When a random initial population of individuals is created, as well as during mutation, the don't care symbol, *, is given a higher probability of selection, in order to ensure rules of varying lengths and candidate knowledge bases of varying sizes. The presence of the don't care symbol also makes the rules more general.

The fitness of an individual rule can be defined as the percentage of instances correctly classified by the rule out of the total number of instances which match the rule's premises. The fitness of a rule indicates the effectiveness of the rule as a discriminator, and the strength of an individual is the sum of the fitness of the rules in that individual. The fitness of an individual is the ratio of its strength divided by the sum of the strengths of all individuals.

Convergence of the genetic algorithm after 20 generations produced a rule set which had a classification accuracy of 75 percent. This was quite good and comparable to the performance of human underwriters. But when a domain expert was given the rules for critique, it was observed that many of the rules did not make sense semantically and some attributes which were not supposed to influence the decision about terminating policy appeared frequently in the rule premise (for example, the attribute type_of_coverage, family_or_individual etc.). These attributes were eliminated from consideration, and the genetic process was run again with the best rule set produced having a classification accuracy of 61 percent, but being more agreeable and in the line of thought of the human expert underwriters.

It is also possible to apply genetic algorithms in order to improve on a rule set derived by another inductive learning technique. As an example the reader is referred to [22] where a hybrid machine learning technique was used for generating the knowledge base. The hybrid approach outperformed both of the other approaches used individually, and is further discussed in Chapter 8.

SUMMARY

Knowledge acquisition is a major bottleneck in the development of expert systems. Several algorithms have been introduced to address this issue and automate the knowledge acquisition process. This chapter examined various machine learning approaches, including derivation of decision trees, version space search type of algorithms and the evolutionary approach. Quinlan's ID3 algorithm follows a divide and conquer approach, and builds classification trees from raw data in a top-down fashion from the root to the leaves. ID3 does not work well for noisy data and it tends to construct classification trees which are too large, rendering the resulting rules too complex to comprehend. Two commonly used approaches to deal with noisy data were presented, namely prepruning and postpruning. The version space search type of algorithm, given next, was able to perform a search through the version space of concepts in order to derive rules. Finally, genetic algorithms and classifier systems were introduced as powerful optimization techniques, and it was shown how they can be used to generate knowledge in the form of rules for an expert system. The connectionist approach to machine learning using neural networks is another very powerful technique and it will be introduced in the next chapter.

EXERCISES

1. Implement the version space search algorithm given in Section 6.5. Choose a domain with available training examples described by a set of attributes, with one attribute designated as the goal attribute. Use the algorithm to derive a rule base which predicts the value of the goal attribute in terms of the values of the other attributes. (To retrieve a wealth of domains and training examples, many of which have been used for benchmark testing of various machine learning algorithms, you can ftp at: *ftp ics.uci.edu* or at *128.195.1.1*)

2. Implement the basic ID3 algorithm and use it to learn a rule base for the same problem domain which was used in Exercise 1.

3. Modify the implementation of ID3, so that it develops the tree using prepruning (or postpruning) and apply it to the same problem domain as in the previous exercises.

4. Develop or obtain an implementation of the basic genetic algorithm in a procedural language (for example, a Pascal implementation is contained in Goldberg's book, ([11]), or surf the Internet to locate other available implementations). Encode the problem domain used in Exercises 1 to 3 so that it is amenable to the evolutionary rule generation approach described in Section 6.6. Find an appropriate fitness function and apply genetic algorithms to generate a rule base.

5. How do the four rule sets developed in the previous exercises compare to each other in terms of simplicity? Accuracy?

REFERENCES

[1] Ash, R., *Information Theory*, Interscience Publishers, New York, 1965.

[2] Breiman, L., Friedman, J., Olshen, R., Stone, C., *Classification and Regression Trees*, Wadsworth, 1984.

[3] Carbonell, J.G., "Derivational analogy: A theory of reconstructive problem solving", in: *Machine Learning II*, Morgan Kaufmann,

1986.

[4] Cestnik, B., Kononenko, I., Bratko, I., "ASSISTANT 86: A knowledge elicitation tool for sophisticated users", *Progress in Machine Learning*, eds. Bratko, I., Lavrac, N., 1987, Sigma Press.

[5] Clark, P., Niblett,T., "Learning If Then Rules in Noisy Domains", *Interactions in Artificial Intelligence and Statistical Methods*, ed. Phelps, B., 1987, Technical Press.

[6] Davis, L., *Handbook of Genetic Algorithms*, Van Nostrand Reinhold, New York, 1991.

[7] Evett, I., Spiehler, E., "Rule induction in forensic sciences", in: *Knowledge Based Systems in Goverment*, Online Pub., Pinner, UK,107-118.

[8] Forsyth, R., "Beagle: A Darwinian approach to pattern recognition", *Kybernetes*, 10, 159-166, 1987.

[9] Forsyth, R., *Machine Learning Principles and Techniques*, Chapman Hall, 1989

[10] Gallant, S., *Neural Network Learning and Expert Systems*, MIT Press, 1993.

[11] Goldberg, D., *Genetic Algorithms in Search, Optimization and Machine Learning*, Addison-Wesley, Reading, MA, 1989.

[12] Greiner, R. "Learning by understanding analogies", *Artif. Intell.*, 35, 81-125, 1988.

[13] Holland, J., Holyoak, K., Nisbett, R., Thagard, P., *Induction: Processes of Inference, Learning and Discovery*, MIT Press, Cambridge, MA, 1986.

[14] Langley, *Scientific Discovery: Explorations of Creative Processes*, MIT Press, 1987.

[15] Kolodner, J., *Case-Based reasoning*, Morgan Kaufmann, 1993.

[16] Michalski, R., Stepp, R., "Learning from observation: conceptual

clustering", *Machine Learning*, Eds. Michalski, R., Carbonell, J., Mitchell, T., Tioga Publishing, 1983.

[17] Mitchell, T.M., "Generalization as search", *Artif. Intell.*, 18, 203-226, 1982.

[18] Mitchell, T.M., Keller, R., Kedar-Cabelli, "Explanation based generalization: A unifying view", *Machine Learning*, 1 (1), 1986.

[19] Natarajan, B., *Machine Learning: A Theoretical Approach*, Morgan Kaufmann, 1991.

[20] Nikolopoulos, C., "A conceptual Clustering and rule generation system", *Proc. Fifth Florida AI Research Symposium*, Ft. Lauderdale, April 1992, 278-284.

[21] Nikolopoulos, C., "BUILT: A Bottom up Inductive Learning Technique", *Proc. of Golden West Intern. Conference on Intelligent Systems*, June 92, Reno, Nevada, 60-66.

[22] Nikolopoulos, C., Duvendack, S., "A Hybrid Machine Learning System and its Application to Insurance Underwriting", *Proceedings of IEEE World Congress on Computational Intelligence*, Evol. Comp. volume, June 1994, 692-695.

[23] Nikolopoulos, C., "Deriving Classification Rules from Neural Nets", *Proceedings of ISIKNH'94*, AAAI/IEEE International Symposium on Integrating Knowledge and Neural Heuristics, May 1994, Pensacola, FL, 141-146.

[24] Quinlan, J.R., "Learning efficient classification procedures and their application to chess end games", in: *Machine Learning: The AI approach*, eds. Michalski, R.S., Carbonell, J., Mitchell, T., Tioga Press, 1983, 463-482.

[25] Quinlan, J.R., "Induction of Decision Trees", *Machine Learning*, 1,81-106, 1986.

[26] Thompson, B., Thompson, W., "Finding rules in Data", *BYTE*, Nov.1986, 149-158.

[27] Weiss, S., Kulikowski, C., *Computer Systems that Learn:*

Classification and prediction Methods from Statistics, Neural Nets, Machine Learning and Expert Systems, Morgan Kaufmann, 1992.

CHAPTER 7

NEURAL NETWORKS

7.1 INTRODUCTION

First generation expert systems are primarily based on the symbolic and logic programming approaches and suffer from the control and knowledge acquisition bottleneck. In second generation expert systems, the symbolic approach is combined with the connectionist and various other learning paradigms to form integrated *hybrid expert systems*.

The connectionist or neural network paradigm provides an alternative to the symbolic approach to building intelligent systems. In some problem domains it may be from difficult to almost impossible to acquire all the necessary knowledge and structure it in terms of logic and knowledge representation constructs. The connectionist approach appears to work quite effectively in such areas where the symbolic approach encounters difficulties (for example: classification, clustering, natural language understanding, learning, vision, and pattern recognition). The general characteristics of these areas are noisy or incomplete input and data, and the inability of human problem solvers to verbalize how they perform the tasks. Problem solving seems to be taking place using knowledge at a subcognitive rather than a cognitive level. Like genetic algorithms, which were described in Chapter 6, neural networks can perform learning and knowledge processing in domains where human expertise is primarily present at a subcognitive level.

In first generation expert systems, the knowledge engineer had to elicit the knowledge from the domain expert, analyze and in a sense predict all the possible cases that can occur in the given domain. In a very complex problem domain, doing so in effect amounts to solving the frame problem and it should come as no surprise that this approach to expert system development leads to the knowledge acquisition bottleneck. Updating the knowledge base, once it is constructed constitutes another problem in expert systems, as trying to detect if inconsistencies are introduced is extremely difficult. Checking the consistency of a knowledge base is an NP-complete problem. In hybrid expert systems on the other hand, part of the knowledge acquisition load can be alleviated as the neural net components of the system are capable of learning themselves from previous experience. It is expected that the incorporation of neural networks and other machine learning techniques in expert systems will render more powerful expert systems, systems that are less prone to reach wrong conclusions in the presence of noisy or partly erroneous data and which will address the knowledge acquisition problem that hinders first generation expert system development.

An introduction to the connectionist model and neural networks is given in this chapter and examples of integrating the symbolic, evolutionary and connectionist paradigms into Hybrid Expert Systems are presented in Chapter 8. For a detailed introduction to the theory and practice of neural computation, the interested reader is referred to ([2, 3, 7, 14, 15]).

7.2 ARTIFICIAL NEURAL NETWORKS

7.2.1 The artificial neuron

An *artificial neuron*, as shown in Figure 7.1, is an autonomous individual processing unit with a set of incoming directed arcs from neighboring neurons and an output signal carried by an outgoing arc which fans out to neighboring cells, (McCulloch and Pitts, [8]). Each arc has a weight associated with it. The weight of the arc from neuron i to neuron j determines the amount of influence neuron i exercises on neuron j. The bigger that weight is, the more related the cells are. Positive weights indicate reinforcement, while negative weights indicate inhibition of one

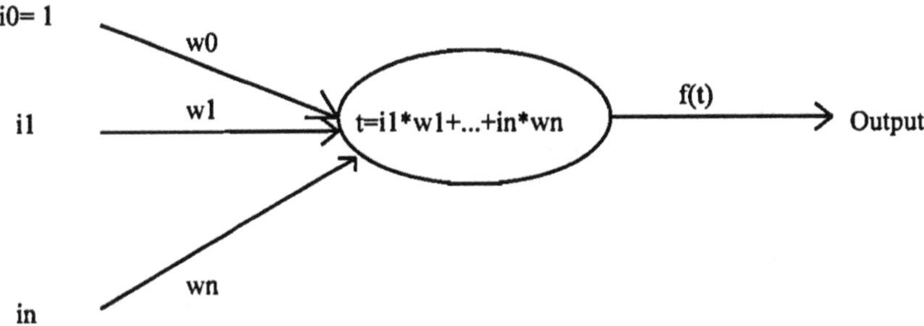

Figure 7.1 An artificial neuron.

concept by the other. The inputs each neuron receives from its neighbors are amplified by the corresponding weights and the total input received determines if a neuron will *fire* (propagate nonzero output to its neighbors through its output arc) or not.

The output of neuron i, also called its *activation value*, can be a discrete value in the set {0,1} or {-1,0,1} or continuous, usually in the interval [0,1] or [-1, +1]. The *total input to neuron i* is calculated by:

$$\sum_{j=1}^{n} out_j * w_{i,j}$$

where out_j is the activation of neuron j and $w_{i,j}$ is the weight of the connection from neuron j to neuron i.

The *threshold (activation, transfer) function* associated with a cell determines what is the output of that cell as a function of its total input. For example, a neuron may fire (produces a nonzero output pulse) when the total input to the cell exceeds a threshold value

$$\sum_{j=1}^{n} out_j * w_{i,j} > \theta$$

where θ is a threshold associated with neuron i.

An *artificial neural network* (ANN) is composed of a number of artificial neurons interconnected in a directed graph structure. An arc connecting cells u_i to u_j has a weight $w_{j,i}$ associated with it. The weight measures the degree of association of the two cells.

According to Rumelhart and McLelland ([12]), a neural network (or a Parallel Distributed Processing (PDP) model, as they called it) is defined by:

1. A set of processing units.
2. The topology of the net, i.e. the pattern of connectivity.
3. An activation function which determines the output of a unit as a function of its total input.
4. A set of weights associated with the links.
5. A learning algorithm, which determines in what way the weights are to be modified in order to improve performance.

One of the advantages of neural nets is their inherent parallelism as each neuron performs its computations locally. This leads to robustness as they can survive processing unit failures and can work even with incomplete or partly erroneous data.

In the sequel, the terms *processing unit, cell or node* will be invariably used in place of the term *neuron.*

7.2.2 ANN topologies

The ANN *topology* or *architecture* is the underlying directed graph structure of units and links. Certain units are designated as input units. An *input* unit takes input from the environment and propagates it to its neighbors. A collection of input units forms the *input layer.* Some units are designated as *output* units and provide the system output. The collection of output units forms the *output layer.* The rest of the units are called *hidden units.*

A *layered network* is one were the nodes are arranged in layers $L_1, L_2, \ldots, L_k$, so that a node in layer i can only be connected to nodes in layer j. A *feed-forward* network is a layered network where there are no directed cycles in the net topology (i.e. j>i). A *feed-backward* or *recurrent* network allows such cycles.

Connectionist models employ various architectures. For example, the Hopfield net model has a bidirectional complete graph architecture with all the cells being input and output cells at the same time. The multilayered perceptron is a feed-forward net with a layer of input units feeding into one or more layers of hidden units, which in turn feed into an output layer.

7.2.3 Activation functions

An activation function determines what is the output of a node given its total input. Usually, units in the same net topology use the same activation function. Examples of transfer functions include the signum function, step function, bidirectional associative memory (BAM), sigmoid, hyperbolic tangent and stochastic functions. Figure 7.2 shows some of these functions.

Consider a net of a single unit using the step function with threshold θ as the activation function. This net is equivalent to the net of

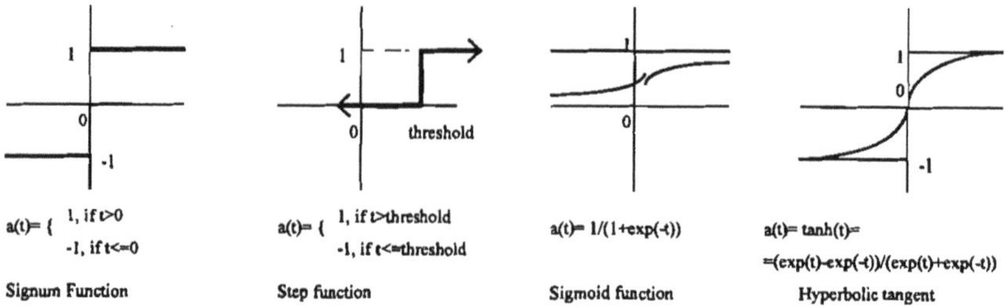

Figure 7.2 Examples of activation functions.

threshold 0 shown in Figure 7.3. An input node was added of constant input equal to one and weight equal to -θ. This node is called a *bias* node. Bias nodes allows us to consider all non-input nodes as having threshold value of zero.

7.2.4 Learning algorithms

A learning algorithm takes as input a set of training vectors, $\{t^k/k=1,...p\}$. Each training vector is of the form $t^k=(t_1^k,...,t_n^k)$, where t_i^k is the input value of input node i and n is the number of input nodes. Learning involves adjusting the network weights until a satisfactory net behavior is reached.

Learning techniques can be classified into two categories: supervised and unsupervised. In *supervised learning*, a training pattern is available together with the desired correct output for that pattern. Each

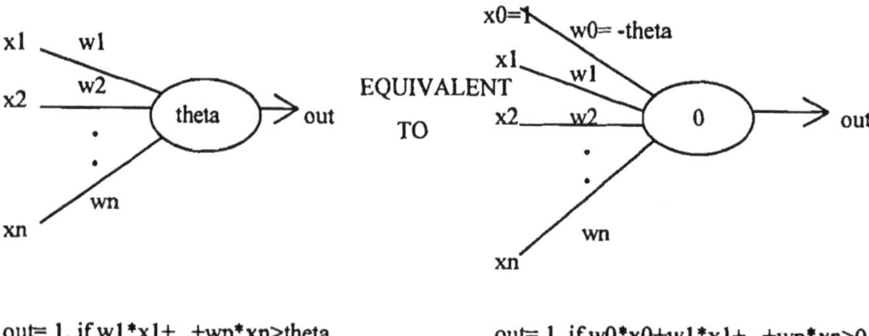

Figure 7.3 Introduction of the bias term.

training pattern will have the form: $t^k=(t_1^k,...,t_n^k,o^k)$, where o^k is the desired output for pattern k. If the net under the current weight set gives a different output, the weights are adjusted to improve the net behavior. In *unsupervised learning*, it is not required that the desired output is supplied with each input pattern. The net evolves by itself until it builds up a set of feature detectors capable of discriminating between sets of input patterns (performs clustering).

A learning algorithm may go through the set of training examples only once or it may repeatedly cycle through them until some predefined criterion evaluating net accuracy is met.

7.3 PERCEPTRON

The single layer perceptron was introduced by Rosenblatt ([10]). A perceptron is shown in Figure 7.4. The input patterns are discrete and assume values in {-1, 0, +1} where 0 indicates that the attribute value is unknown. The bias node has weight w_0. The activation function for each neuron is:

$$\text{output} = \begin{cases} 1, \text{if } \Sigma_{j=0}^{n} w_i x_i > 0 \\ 0, \text{otherwise} \end{cases}$$

Given a set of training examples, each containing a data pattern and a corresponding desired output for that pattern, we want to compute the set of weights $\{w_0, w_1, ..., w_n\}$ so that the net gives correct responses for the input patterns. For an algorithm to achieve this (the perceptron's learning algorithm), the reader is referred to Rosenblatt ([10]).

There are some classes of problems that the single layer perceptron cannot learn to solve. For example, simulating the logical function exclusive-or (XOR) or determining if a given object is surrounded by another object are tasks that cannot be learned by a single layer perceptron. Multilayered perceptrons can solve such problems, but the perceptron learning algorithm can not be used to effectively learn these problems. In their book "Perceptrons", Minsky and Papert, ([9]), further postulated that there can be no learning algorithm to train a multilayered perceptron to solve such problems. This undermined the validity of neural net research and led to the abandonment of the field by

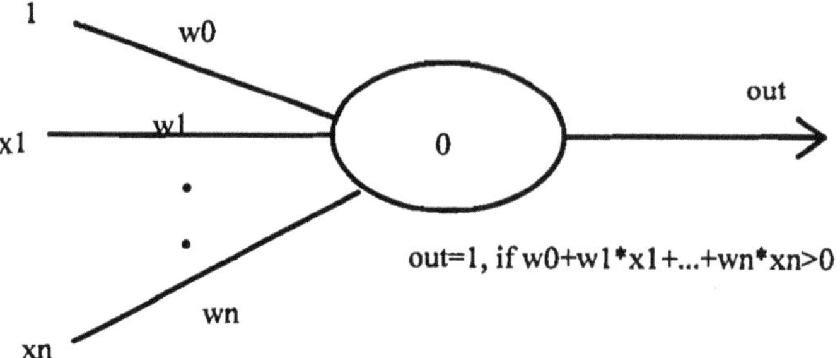

Figure 7.4 A perceptron.

many researchers. In the early 80s, the major breakthroughs of Hopfield nets, Boltzmann machines and backpropagation brought back neural nets as a major contributor to intelligent systems research.

The type of problems a perceptron can solve are described as follows.

Perceptron training patterns with coordinates equal to -1, 0 or +1 can be viewed as points in n-dimensional Euclidean space. A set of patterns is called *separable* iff there is a n-1 dimensional hyperplane (called a *decision surface*) separating the group of points for patterns whose output is +1 from the group of points for patterns whose output is -1. It was shown that if the training set of patterns is *separable* then the perceptron learning algorithm converges to a weight set which classifies all input patterns correctly. For example, as shown in Figure 7.5, the patterns for the logical AND are separable, since there is a line (hyperplane of dimension 1) separating them and a perceptron can be trained to simulate the logical AND, but those corresponding to XOR are not separable. Points corresponding to patterns whose output is supposed to be +1 (true) are represented in the figure by filled in circles and the ones corresponding to patterns with output -1 (false) by circles. If the set of training patterns is nonseparable then the perceptron learning algorithm diverges even though the weight set does not approach infinity but remains bounded (Minsky and Papert [9]).

We give below a brief justification of the fact that separability of patterns is necessary using as example a perceptron of two input nodes and one output node.

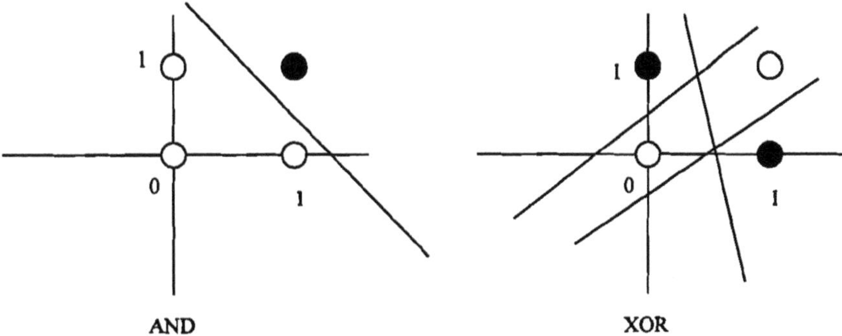

Figure 7.5 Separable and non-separable pattern sets.

The total input to the output node is equal to:
$I=w_0+w_1x_1+w_2x_2$.

If I=0, solving the equation for x_2 yields:
$x_2= -(w_1/w_2)x_1- (w_0/w_2)$
or
$f(x_1,x_2)=x_2+(w_1/w_2)x_1+(w_0/w_2)=0$

For a fixed set of weights, this equation represents a straight line in the 2-dimensional plane. Input vectors $(1, x_1, x_2)$ on one of the half planes defined by the straight line will make $f(x_1,x_2)$ positive and give output 1 and vectors on the other half plane will make it negative and give output -1. Therefore, a set of training patterns can be correctly classified by a single layer perceptron iff a weight set can be found which defines a line (hyperplane in general) separating the input vectors with desired output 1 from the ones with desired output 0.

7.4 HIDDEN LAYERS, MULTILAYER PERCEPTRONS

If more layers of nodes are introduced between the input and output layers then these multilayer perceptrons can solve nonseparable problems. For example, the net of Figure 7.6 simulates the XOR problem. The effect of multilayered perceptrons is that multiple hyperplanes are drawn to separate the patterns with +1 desired output from the ones with -1 desired output. For example, the net of Figure 7.6 draws line (1) using the hidden node and the output node draws line (2). The intersection of the two half

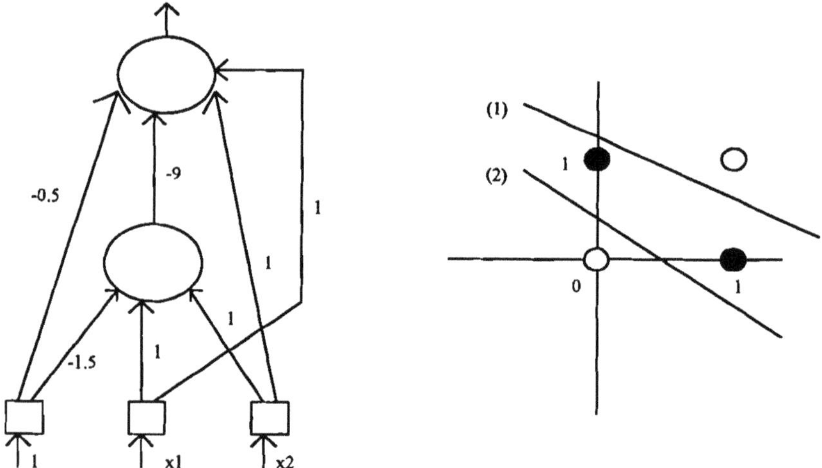

Figure 7.6 A multilayered perceptron for the XOR problem.

planes separates the two sets of points.

A perceptron with two hidden layers can draw any convex rectangular region in order to separate the two sets of points. A perceptron with three hidden layers gives us the ability to separate any finite set of points, as any convex or nonconvex region can be drawn. No more than three hidden layers are needed to solve a problem. The perceptron learning algorithm as presented by Rosenblatt, ([10]), of course does not work for training multilayered nets as the desired output of the processing units in the hidden layer is not known. The introduction of backpropagation by Rumelhart, Hinton and Williams, ([11]), was a major breakthrough as it solved this problem of how to train multilayered nets. Backpropagation is presented in Section 7.6.

7.5 HOPFIELD NETWORKS

The Hopfield net is an associative memory in the sense that it can recall stored patterns upon being presented with corrupted or noisy input patterns. A Hopfield net of N nodes has the ability to memorize and recall perfectly a maximum of N/4logN patterns. Hopfield nets can also be used to solve optimization problems.

The topology of the Hopfield net is a complete graph with each cell connected to all other cells. The weights can be organized in the form of an n by n matrix, where n is the number of nodes in the net and $w_{i,j}$ represents the weight from node j to node i. The weight matrix is symmetric, as $w_{i,j}=w_{j,i}$. Self arcs are not allowed, so $w_{i,i}=0$.

The activation value of node i is given by:

$$a(i) = \begin{cases} +1, \text{ if } & h_i >= 0 \\ -1, \text{ otherwise} & \end{cases}$$

where $h_i=\Sigma_j w_{i,j} a(j)$ is the total input to node i.

The input patterns to be memorized by the net are vectors of dimension n, where n is the number of neurons in the net. The vector coordinates assume the values +1 or -1. The ith coordinate of an input vector is assigned as the input and initial activation of the ith neuron. The activation function employed by neurons is a step function giving output +1 if the total input is positive, -1 otherwise.

After the initial activations are imposed on the neurons, the net will start recalculating its activations in cycles until a stable state is reached. A *stable state* is one where the activations of the nodes remain unchanged. The output of the net is the vector of activation values of the neurons in that stable state. The assignment of the weights determines which patterns the net has memorized. The update of node activations can be done *asynchronously*, where nodes are updated one at a time (the order in which updating is done is not important) or *synchronously*, where nodes are updated in parallel. The learning algorithm which will be described below, chooses the weights in such a way that the stable states correspond to the stored patterns that we wish the net to memorize. These stored patterns are called *attractors*.

An energy function is also associated with a Hopfield net which is a function of the activation state of the net. The attractors correspond to minima of the energy function. When an input pattern is presented to the net, the net evolves towards the nearest attractor.

7.5.1 Hopfield learning algorithm

The learning rule used by Hopfield nets is a Hebb rule. It is analogous to associative learning in humans, where a cognitive association is created between two concepts by repeated presentation of simultaneous stimuli. A Hebbian rule basically states that if two neurons are simultaneously excited then the synaptic strength between them has to be increased.

Let $x^{(1)}, x^{(2)}, \ldots, x^{(k)}$ be the k patterns we want the net to memorize, where for each j: $x^{(j)}=(x^{(j)}_1, \ldots, x^{(j)}_n)$ is a vector of dimension n. The number of neurons in the net is n and the value of each pattern coordinate is either +1 or -1.
The weight between neuron i and neuron j is computed by:

$$w_{i,j} = \begin{cases} (1/n) \sum_{s=1}^{k} x_i^s \, x_j^s, & \text{if i is not equal to j} \\ 0, & \text{if } i=j \end{cases}$$

7.5.2 Net evolution

The updating of node activations proceeds as follows:

Consider the input pattern ($input_1$, ..., $input_n$) where for each i: $input_i=\pm 1$. In each cycle, the activations of all nodes are calculated until they become stable from cycle to cycle.

1. Let $t=0$. Initialize the activation of each node as: $out_i(t)=input_i$, $\forall i=1,\ldots,n$.
2. Repeat
For each i: $1\leq i \leq n$ Let $out_i(t+1)=1$, if $\sum_{j=1}^{n} w_{i,j}\, out_j(t)\geq 0$
or $out_i(t+1)=-1$, otherwise
Let $t=t+1$
Until $out_i(t)=out_i(t-1)$ $\forall i$: $1\leq i \leq n$.
3. The stable pattern ($out_1(t)$, $out_2(t)$, ..., $out_n(t)$) is the net output.

7.5.3 Energy Function

Hopfield introduced the idea of an *energy function*, which, given a state of activations, calculates a corresponding energy value for the system. The learning algorithm chooses the weights in such a way that the stored attractors are states which minimize the energy function. The energy function is a type of Lyapunov function.

The *energy function* is defined as follows:

$$E(a_1, a_2, \ldots, a_n) = -(1/2) \Sigma_{i,j} w_{i,j} a_i a_j$$

where a_i is the activation of cell i.

Given an initial configuration of cell activations, the values of the energy function decrease as the system evolves. The system will converge and stabilize to a set of activations which corresponds to a local minimum of the energy function. A problem with Hopfield nets using a step transfer function and discrete outputs is that they will not always converge to the stored attractor closest to the input pattern. The pattern on which they stabilize may be none of the attractors. These other stable states correspond to local minima of the energy function and can be classified into *inverse states* (exactly opposite of a memorized attractor), *mixture states* (linear combinations of memorized attractors) and *spin glass states* (not related to any combination of attractors). The inverse states are easy to detect but the other two, called *spurious minima*, are more troublesome.

7.5.4 Stochastic Hopfield networks

A solution to the spurious minima problem is to introduce stochastic Hopfield nets, ([5]), where each node uses a logistic transfer function instead of a hard limiter step function. Boltzmann machines ([1]) are similar to stochastic Hopfield nets, but unlike Hopfield nets they also have a layer of hidden nodes.

The activation a_i of node i is probabilistically defined as:

$$a_i = \begin{cases} +1, \text{ with probability } P_i(+1) = 1/(1+e^{-2h_i/T}) \\ 0, \text{ with probability } P_i(0) = 1 - P_i(+1) = 1/(1+e^{2h_i/T}) \end{cases}$$

where h_i is the total input to node i : $h_i = \Sigma_j w_{i,j} a_j$ and T is a variable used to introduce one more degree of freedom into the system.

The logistic function introduced above was used in the Glauber dynamics theory from physics, which extends the Ising model by explaining the influence of temperature on properties of magnetic materials, ([4]). In the physics model, T represented the environment temperature. In the stochastic Hopfield model T is still called *temperature* even though it does not correspond to any physical concept. The value of the variable T is used to help get the system out of stagnation at a local minimum of the energy function. If the Hopfield net stabilizes in an unrecognizable state (spurious minima of the energy function), the value of T is changed and activations recalculated, in effect kicking the system out of the local minimum it had settled on. There is no known methodology for picking what the value of T should be. A value which is too high may cause a big step change in the energy function value with the result of skipping over and completely missing the closest local minimum. A value which is too small may not be sufficient to get the system out of the stagnation in a spurious minimum. In practice, the value of T is usually started higher and gradually decreased as the model evolves. Among the weaknesses of stochastic Hopfield nets and Boltzmann machines is that they are very labor intensive and slow.

It is worth noting that, Hopfield nets with continuous activation values have also been introduced, where the activation function is:

$a(i)=\tanh(h_i/T)$, the hyperbolic tangent giving cell output in the range [-1,+1]

or

$a(i)=1/(1+e^{-2hi/T})$, the sigmoid function giving output in the range [0,1].

7.5.5 Solving optimization problems with Hopfield nets

The concept of the energy function for Hopfield nets can be used to approximate solutions to NP-complete problems (problems for which no polynomial time algorithm is known). We give below an example of such a use in the TSP (traveling salesman problem).

Solving optimization problems through the use of Hopfield nets involves the following steps:

1. Encode the problem into a net: Determine the net topology, what each cell activation means in term of the problem concepts, how

a state of node activations corresponds to a problem solution.

2. Configure the quantity which needs to be optimized as a function of cell activations.

3. Describe any constraints that have to be imposed on a net activation state if it is to correspond to a legitimate problem solution.

4. Define the energy function as a function of what is to be optimized and penalty terms imposed for patterns which violate the constraints.

5. Manipulate the energy function algebraically in order to reduce it to a quadratic formula in terms of the cell activations:
$\text{Energy} = \Sigma_{i,j}\ \text{coefficient}_1(i,j)*a_i a_j + \Sigma_{i,j}\ \text{coefficient}_2(i)*a_i$
+(constant term without activations).

The coefficient of a quadratic term $a_i a_j$ is the weight of the link between nodes i and j. The coefficient of the linear term a_i is the threshold of node i. For this choice of weights, it can be proven that the minima of the energy function are the net attractors and they correspond to approximate solutions to the optimization problem represented by the net.

To exhibit this technique, we show its application to the TSP problem (see Hopfield and Tank, ([6]), Hertz, Krogh, Palmer, ([4])).

The TSP problem is stated as: *Given N cities and the distances between them find a closed tour of the smallest total length.* A closed tour is a path which visits every node in a graph exactly once before returning to the starting node.

A stochastic Hopfield net of n^2 binary nodes was used, where n is the number of cities. The units are labeled $unit_{1,1}$, $unit_{1,2}$, ..., $unit_{1,n}$, ..., $unit_{n,1}$, ..., $unit_{n,n}$. Let $a_{i,j}$ represent the activation of $unit_{i,j}$. A state of node activations is interpreted as follows. Node $unit_{i,j}$ has activation equal to 1 ($a_{i,j}=1$) if city i is in the j-th stop on the tour, and activation 0 otherwise. For example, the net of Figure 7.7 corresponds to five cities. In the figure, nodes with activation equal to 1 are denoted by a shaded circle. Each unit is connected to all other units. The links are not shown to avoid clustering. The activation pattern shown on the net corresponds to the legal tour 2,4,1,5,3,2.

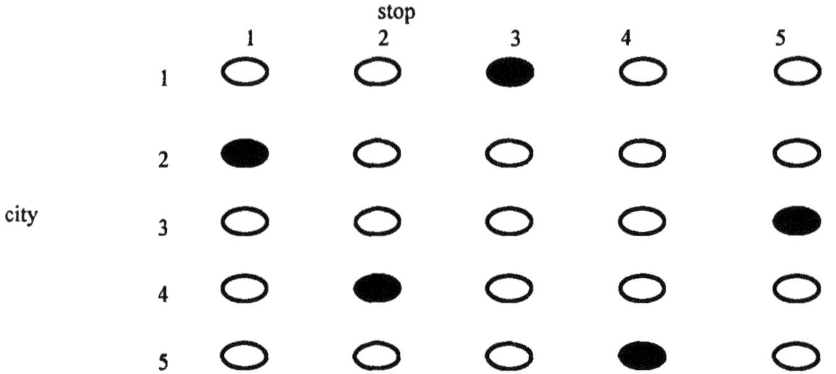

Figure 7.7 A Hopfield net topology for the TSP.

For an activation pattern to be a legal tour, each column of cells must have exactly one node of activation 1 as only one city can be visited in a stop. Similarly each row must have exactly one cell of activation 1, as we are not allowed to visit a city twice. These constraints on what constitutes a legal tour are expressed in equation form as follows:

$\forall i \; \sum_j a_{i,j}=1$ and $\forall j \; \sum_i a_{i,j}=1$

If $d_{i,j}$ is the distance between city i and city j, then the total length of a tour can be expressed as:

$$\text{Length}=(1/2)\sum_{i,j} d_{ij} a_{i,k}(a_{j,k+1}+a_{j,k-1})$$

Finally, the energy function is expressed as the sum of what is to be minimized and penalty terms for each constraint. If the constraint is satisfied by the state of activations then the penalty term should be equal to 0. In the TSP problem the energy function then becomes:

$$\text{Energy}=(1/2)\sum_{i,j,k} d_{i,j} a_{i,k}(a_{j,k+1}+a_{j,k-1})+(\eta/2)[\sum_k(1-\sum_i a_{i,k})^2+\sum_i(1-\sum_k a_{i,k})^2]$$

By developing the squares and algebraically expanding we get:

$$\text{Energy}=(1/2)\sum_{i,j,k} d_{i,j} a_{i,k} a_{j,k+1} + (1/2)\sum_{i,j,k} d_{i,j} a_{i,k} a_{j,k-1} - \eta \sum_{i \neq j} a_{i,k} a_{j,k} - \eta \sum_{k \neq l} a_{i,k} a_{i,l} - \eta \sum_{i,k} a_{i,k} + \eta N$$

The coefficients of the quadratic terms are the weights. So, the weight $w_{(i,j),(k,l)}$ between nodes $unit_{i,j}$ and $unit_{k,l}$ is equal to η for nodes on the same row or column and is equal to $-d_{ik}$ otherwise. The coefficients of the linear terms are the node thresholds, so each node has a threshold of $-\eta$.

Using the matrix topology and the weights given above, a stochastic Hopfield net will stabilize to a state of activations representing a near optimal tour.

7.6 MULTILAYERED NETS AND BACKPROPAGATION

The introduction of the backpropagation learning algorithm played a major role in the rebirth of neural networks as a viable computational model. Backpropagation effectively trains feed-forward multilayered networks to solve the non-linearly separable problems which the perceptron learning algorithm fails to solve.

The cells of a multilayered net trained by backpropagation employ a non-linear and continuous activation function. The activation values can be in the interval [0,1] or [-1,+1]. Backpropagation is a gradient descent based algorithm and it adjusts the weights in order to reduce the mean square error between desired and actual output.

The total input to node i is given by $h_i = \sum_k w_{i,k} a_k$, where a_k is the activation of node k. The activation function is given by one of the following functions depending on the desired range for the cell output.

$$\text{activation of node } i = a_i = \begin{cases} 1/(1+e^{-2h_i}), & \text{for node output in the range } [0,1] \\ \tanh(h_i), & \text{for output in the range } [-1, +1] \end{cases}$$

Assume a net of n input nodes labeled $input_1$ to $input_n$, h hidden nodes labeled $hidden_1$ to $hidden_h$, and m output nodes labeled $output_1$ to $output_m$. Consider the set of training patterns $x^1, x^2, \ldots, x^p$ where each x^i is a vector of dimension n+m. The first n coordinates is the input pattern and the last m coordinates $x^i_{n+1}, x^i_{n+2}, \ldots, x^i_{n+m}$ are the desired outputs for input pattern i from each of the output nodes.

Given a set of weights W, the mean square error between actual and desired output is defined as the sum of the squares of the output differences between desired and real output :

$$\mathrm{Error}(W)=(1/2)\ \Sigma_{k,i}[d^k_i-r^k_i]^2 \qquad (1)$$

where d^k_i is the desired output from output node i for input pattern k, and r^k_i is the real output from output node i when the k-th input pattern is fed into the net. The error function is a function of the weights since different weights will produce different real outputs and therefore different errors. The error is always greater or equal to 0 and when it is 0 the net behaves correctly on all p input patterns.

To find the weight set resulting in a global minimum for the error function, a gradient descent method is applied where the weights are changed in the direction of the partial derivatives of the error function and in an amount proportional to it, i.e. $\Delta w_{i,j} = -\eta\ (d\ E(W)/d\ w_{i,j})$. The variable η, called the *learning rate*, determines the amount of change and its value is between 0 and 1. The error is calculated at the output nodes and then propagated backwards to adjust the weights. In *single pattern training* after each input pattern is presented the weights are adjusted by back propagation, while in *epoch training* the whole set of input patterns is presented, the error is accumulated and then back propagation takes place. Practice has shown that epoch training usually converges to a good weight set faster than single pattern training.

For $w_{i,j}$ the weight from a node j in the last hidden layer to an output node i, partial differentiation of the error function gives

$$\Delta w_{i,j}= \eta\Sigma_p\ \delta^p_i\ r^p_j \qquad (2)$$

where $\delta^p_i=(d^p_i-r^p_i)r^p_i(1-r^p_i)$ and p ranges over the input patterns.

The new weight after presentation of the training set at the t-th cycle (epoch) is calculated from the old weight in epoch t-1 by:

$$w_{i,j}(t)=w_{i,j}(t-1)+\Delta w_{i,j}$$

For weights between two hidden nodes or between an input and a hidden node, the formula above is not applicable since the desired output of a hidden node is not known. For such weights, the following formula is used instead:

$$w_{i,j}(t)=w_{i,j}(t-1)+\eta\Sigma_p\delta^p_j r^p_i \quad (3)$$

where $\delta^p_j=r^p_j(1-r^p_j)\Sigma_k w_{k,j}\delta^p_k$ and p ranges over the input pattern set.

7.6.1 Backpropagation learning algorithm

The backpropagation algorithm is as follows:

Randomly initialize the weights.
Repeat
For each training pattern p compute the real output r^p_j from each node j in the output layer.
Compute the mean square output error using formula (1).
Modify the weights going backwards from output layer towards the input layer. For the links between hidden nodes and output nodes use formula (2). For the rest of the links use formula (3).
Until (the mean square error is less than a predefined value or maximum number of iterations has been reached).

In order to improve the performance of the algorithm sometimes a momentum term, which is a fraction of the previous weight change, is added to the equation:

$$w_{i,j}(t)=w_{i,j}(t-1)+\eta\Delta w_{i,j}(t)+\alpha\Delta w_{i,j}(t-1)$$

The *momentum coefficient* α is chosen between 0 and 1 and the value 0.9 seems to work reasonably well in many applications.

7.6.2 An example on backpropagation

Consider the net topology of Figure 7.8 which is to be trained to perform the logical *exclusive or* (XOR) function. The training patterns are $p_1=(0,0,0)$, $p_2=(0,1,1)$, $p_3=(1,0,1)$, $p_4=(1,1,0)$ where the third coordinate is the desired output. The threshold function in use is $f(x)=1/(1+e^{-x})$.

The weights are randomly initialized as shown in Figure 7.8. We

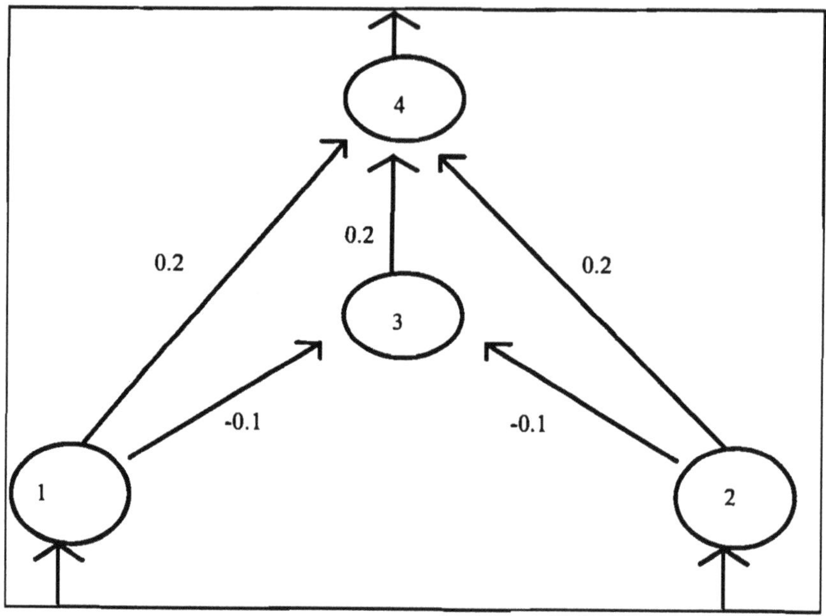

Figure 7.8 Topology and initial weights for the backpropagation XOR net.

calculate the original mean square error for these weights and then apply the back propagation algorithm for one epoch to configure the new weights. The mean square error indeed is reduced with the new weight set.

For input pattern p_1 the real output is $1/(1+e^{-0.2})=0.5498$.
For input pattern p_2 the real output is $1/[1+e^{-(0.2/(1+\exp(0.1))+0.2)}]=0.5732$.
For input pattern p_3 the real output is the same as for p_2.
For input pattern p_4 the real output is $1/[1+e^{-(0.2+0.2+(0.2/(1+\exp(0.2)))}]=0.6201$.
Therefore the total mean square error for the given weight set is:

$$\text{Error}=1/2\Sigma_p(d^p{}_4-r^p{}_4)^2=$$
$$1/2[(0-0.5498)^2+(1-0.5732)^2+(1-0.5732)^2+(0-0.6201)^2]=1.0511$$

Suppose the value of the learning rate η is 0.5. The weights to the output layer node are adjusted according to the formula:

$$w_{4,j}(1)=w_{4,j}(0)+\eta\Sigma_{p=1,4}(d^p{}_4-r^p{}_4)r^p{}_4(1-r^p{}_4)$$
to get:

$w_{4,1}$=0.2+0.5[(0-0.5498)(0.5498)(1-0.5498)+2(1-0.5732)(0.5732)(1-0.5732)+(0-0.6201)(0.6201)(1-0.6201)]=0.16=$w_{4,2}$=$w_{4,3}$

The weights not leading to output layers are updated by:

$w_{i,j}(1)=w_{i,j}(0)+\eta\Sigma_p([r^p_j(1-r^p_j)\Sigma_k w_{k,j}\delta_k])r^p_i)$
so,
$w_{3,1}$=-0.1 and $w_{3,2}$=-0.1.

Under the new weights the output for pattern p_1 is 0.5399, for pattern p_2 is 0.5587, for pattern p_3 is 0.5587 and pattern p_4 0.5967. The new mean square error is 1.0370. After the first cycle the error was reduced by 1.0511-1.0370=0.0141. The process is repeated until the error is reduced to some acceptable level.

SUMMARY

In this chapter, two of the most widely used connectionist models, the Hopfield net and multilayered feed-forward net trained with back propagation, were introduced. Neural Networks are not good at symbolic computations, but they are good at tasks experts perform at the subcognitive level, such as pattern recognition, classification, clustering and handling distorted data. The parallel architecture of neural nets can give them increased speed and enables improved fault tolerance, robustness and graceful degradation. They can overcome partially correct or incomplete data. Neural nets do not suffer from the knowledge acquisition bottleneck which is a main difficulty in the development of logic based expert systems. The two technologies can help each other in providing more robust and effective expert systems. Hybrid expert systems combining the connectionist and symbolic paradigms are examined in the next chapter.

EXERCISES

1*. Implement a Hopfield neural network model of 9 by 13 nodes in a procedural language or Prolog. Use it to memorize 2 characters represented in a bit vector representation in a 9 by 13 bit map. Test its ability to recognize corrupted versions of these patterns. Repeat trying to memorize 3, 4, 5 and 6 patterns. What is the maximum number of patterns the net can memorize? How would you modify the net if you need to memorize more patterns?

2*. Implement the backpropagation algorithm for a feed-forward net using a procedural language or Prolog. Train a net to classify the loan approval knowledge base of Section 5.2. Propose another classification problem for which you have available training instances and train a neural net to solve it...(a good supply of raw data for experimentation is provided at the ftp ics.uci.edu).

REFERENCES

[1] Ackley, D.H., Hinton, G.E., Sejnowski, T.J., "A Learning Algorithm for Boltzmann Machines", *Cognitive Science*, 9, 1985, 147-169.

[2] Gallant, S., *Neural Network Learning and Expert Systems*, MIT Press, 1993.

[3] Hecht-Nielsen, R., "Neurocomputer Applications", in: *Neural Computers*, ed. Eckmiller, R. and Malsburg, V.D., Springer Verlag, 1989.

[4] Hertz, J., Krogh, A., Palmer, R., *Introduction to the theory of Neural Computation*, Addison Wesley, 1991.

[5] Hopfield, J.J., "Neural Networks and Physical Systems with Emergent Collective Computational Abilities", *Proceedings of the National Academy of Sciences*, 79, 1982, 2554-2558.

[6] Hopfield, J.J., Tank, D.W., "Neural Computation of Decisions in Optimization Problems", *Biological Cybernetics*, 52, 1985, 141-152.

[7] Lippmann, R.P., "An Introduction to Computing with Neural Nets", *IEEE ASSP Magazine*, April 1987.

[8] McCulloch, W.S., Pitts, W. "A logical calculus of ideas immanent in nervous activity", *Bulletin of Mathematical Biophysics*, 5, 115-133.

[9] Minsky, M. Papert, S., *Perceptrons*, MIT Press, Cambridge, 1969.

[10] Rosenblatt, F., *Principles of Neurodynamics: Perceptrons and the theory of Brain Mechanisms*, Spartan Press, Washington, D.C., 1961.

[11] Rumelhart, D.E., Hinton, G.E., Williams, R.J., "Learning Internal Representations by Error Propagation", in: Rumelhart, D.E. and McClelland, J.: *Parallel Distributed Processing: Explorations in the Microstructures of Cognition*, Vol 1, MIT Press, 1986.

[12] Rumelhart, D.E., McClelland, J., *Parallel Distributed Processing: Explorations in Microstructure of Cognition*, Vol. I, II, MIT Press, Cambridge, MA, 1986.

[13] Sejnowski, T.J., Rosenberg, C.R., "NETtalk: A Parallel Network that Learns to Read Aloud", John Hopkins Univ., Dept. of EE and CS, *Tech. report* 86/01/1986.

[14] Sejnowski, T.J., "Neural Network Learning Algorithms", in: *Neural Computers*, ed. Eckmiller, R. and Malsburg, V.D., Springer Verlag, 1989, 291-300.

[15] Windrow, R., Lehr, M.A., "30 Years of Adaptive Neural Networks: Perceptron, Madaline and Back Propagation", *Proc. of IEEE*, Vol. 78, No. 9, Sept. 1990, 1415-1442.

CHAPTER 8

HYBRID EXPERT SYSTEMS

8.1 INTRODUCTION

In addition to the hybrid knowledge representation schemes discussed in Chapter 3 or hybrid inference techniques, as discussed in Chapter 2, a new breed of hybrid expert systems has emerged. Hybrid systems combine the symbolic, connectionist and evolutionary paradigms into a unified system while retaining many of the advantages of each paradigm in order to form more powerful and efficient expert systems. This chapter illustrates how these seemingly unrelated paradigms can work together and complement each other.

Three case studies are presented. The first describes connectionist expert systems (exemplified by the system MACIE) which are neural inference nets capable of providing explanations based on an implicitly stored knowledge base of rules.

Next, an investment advising expert system is described which combines the rule based symbolic approach with the neural network paradigm in two independent but interfacing modules within the same system. The two modules are used to solve different subtasks in the problem domain and communicate with each other to pass results or data to each other. A method of automating the process of neural net configuration by using genetic algorithms and classifier systems was used in the system. Any feed-forward neural net is equivalent to a classifier system and a method exists for performing a conversion from a neural net to a classifier system and vise versa. Genetic Algorithms can then be applied to a population of classifier systems, in order to find a near optimal neural net architecture which learns a given set of training patterns. The algorithms used have been implemented in the system Neutogen, (NEUral net TOpology GENerator, [11]). Neutogen configures the total network architecture including the topology and weight matrix of the net.

Finally an expert system for insurance underwriting is presented which uses machine learning techniques and the evolutionary approach to automatically generate the knowledge base for the problem domain bypassing the traditional knowledge acquisition process. For additional applications and examples of hybrid systems in a number of domains, the reader is referred to [2,4,5,14].

8.2 MACIE: A CONNECTIONIST EXPERT SYSTEM

MACIE (Matrix Controlled Inference Engine) is a typical example of a *neural expert system* (Gallant, [2]). The principles and algorithms underlying MACIE form the basis of *Knowledge Net,* a commercial product developed by HNC Inc.

A MACIE neural network is shown in Figure 8.1. The nodes of the net are labeled with the concepts which they represent. The existence of a link between cells c_i and c_j indicates that there is a dependency relation from cell i to cell j (support or inhibition depending on the sign of the weight). When the net architecture is constructed such known conceptual dependencies are taken into account. The net could also be completely connected in the absence of sufficient knowledge about dependencies. The net is trained by using a set of training examples to calculate the appropriate weights.

The activation function is given by:

$$S_i = sign(\textstyle\sum_j w_{ij} S_j),$$

where S_j is the activation of node j and w_{ij} is the weight of the link from node j to node i.

The input layer nodes can be assigned the values of +1 (true), -1 (false) or 0 (do not know). It should be noted that the methodology described in this section can also be extended to networks of continuous inputs/outputs (Gallant, [2]).

Each node in the network has a question associated with it, which typically inquires about the value of the concept corresponding to that node. For example, the question "Does it have language skills?" may be assigned to node *language_skills*. Whenever the output of a node n cannot be determined due to insufficient input data, and m is a predecessor of node n with an undetermined value for the corresponding concept, then the question associated with the predecessor node m is posed to the user. Backward tracing of the net takes place starting at the output node and

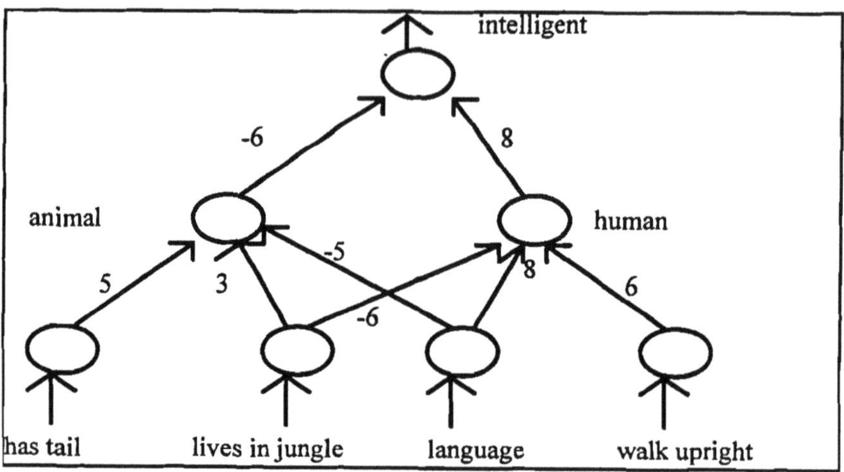

Figure 8.1 A MACIE architecture.

asking the questions related to the immediate predecessors. Only the number of inputs which is sufficient to determine the output of a node is sought from the user. When the given partial input to a node is enough to determine the value of the output, no further questioning takes place. Instead forward chaining takes place propagating the output through the net.

In order to illustrate this process, consider the network of Figure 8.1 and assume that the following questions are attached to the concept nodes:

node: ANIMAL	Question: Is it an animal?
node: HUMAN	Question: Is it a human?
node: TAIL	Question: Does it have a tail?
node: LIVES_IN_JUNGLE	Question: Does it live in the jungle?
node: LANGUAGE	Question: Does it have language skills?
node: WALKS	Question: Does it walk upright?

A run of MACIE on this network knowledge base will produce the following (system prompts are in upper case).

RUN MACIE
INPUT INITIAL VALUES FOR INPUT VARIABLES:
tail -1

language +1
walks +1
PURSUING: INTELLIGENT
PURSUING: ANIMAL
PURSUING: HUMAN
CONCLUDE: ANIMAL IS FALSE
CONCLUDE: HUMAN IS TRUE
CONCLUDE: INTELLIGENT IS TRUE

In this case, the given input was enough to determine the output of nodes ANIMAL and HUMAN and therefore of node INTELLIGENT. The total input to node ANIMAL from nodes TAIL and LANGUAGE is equal to:

$$(-1)(5)+(+1)(-5)= -10<0.$$

The value of the third input node LIVES_IN_JUNGLE is unknown, but there is no need to ask for its value. The total input to node ANIMAL will still be negative independent of the value of LIVES_IN_JUNGLE. Similarly, the total input to node HUMAN is equal to:

$$(1)(8)+(1)(6)=14 > 0.$$

The value of input variable LIVES_IN_JUNGLE is unknown but no matter what its value is, it would have no effect on the sign of the total input. As a result, HUMAN is also determined to be true and forward chaining takes place to yield INTELLIGENT is true.

In general, MACIE performs the following inference test. Let c be a node which receives input from nodes $c_1,\ldots,c_n$ with corresponding weights $w_1,\ldots,w_n$. Let $N=\{1,\ldots,n\}$ and $I=\{i_1,\ldots,i_k\}\subseteq N$ be the set of indexes of nodes whose output value is known.

The total input to node c from its neighbors of known values is:

$$\text{RELEVANT}=\textstyle\sum_j S_j w_j,$$

where $j \in I$ and S_j is the activation of node j.

The total potential contribution from the nodes whose values are not known is in the range ±POTENTIAL where:

POTENTIAL=$\sum_k |w_k|$, with $k \in$ N-I.

The output of node c is determined regardless of what are the values assumed by the unknown input nodes iff |RELEVANT| > POTENTIAL. If this condition is violated then the next node from the set N-I is picked and an attempt is made to determine its value. If it is a concept node the question is posed to the user. If the answer is *unknown* then backward chaining takes place in order to derive the unknown value from its predecessors.

A second example is given below in order to further illustrate these points.

```
RUN MACIE
INPUT INITIAL VALUES FOR INPUT VARIABLES:
language +1
PURSUING: INTELLIGENT
PURSUING: ANIMAL
IS IT AN ANIMAL?
> y)es, n)o, u)known, ?)explain
u
DOES IT HAVE A TAIL?
 > y)es, n)o, u)known, ?)explain
n
CONCLUDE: ANIMAL FALSE
PURSUING: HUMAN
IS IT A HUMAN?
 > y)es, n)o, u)known, ?)explain
u
DOES IT LIVE IN THE JUNGLE?
> y)es, n)o, u)known, ?)explain
n
CONCLUDE: HUMAN IS TRUE
CONCLUDE: INTELLIGENT IS TRUE
TRY A NEW CASE?
>y)es, n)o, ?)explain
?
WHICH VARIABLE?
animal
ANIMAL IS FALSE DUE TO THE FOLLOWING RULE:
IF LANGUAGE IS TRUE AND TAIL IS FALSE THEN ANIMAL IS
FALSE
```

The given input of +1 for the language input node is not enough to determine the output of node animal, since

RELEVANT(animal) = (+1)(-5)=-5 and
|RELEVANT(animal)| ≤ POTENTIAL(animal)=5+3=8.

Similarly, knowing the language input node is not sufficient for determining the value of "human", therefore the value of "intelligent". As a result, backtracking takes place from the net output to the node animal and the associated question is posed to the user. The answer "unknown" is given, so the question attached to the node *has_tail* is posed. Given the negative answer, the calculations

RELEVANT(animal)=(-1)(5)+(+1)(-5)= -10

and

POTENTIAL(animal)=3

are performed and since |RELEVANT(animal)| is greater than POTENTIAL(animal), the conclusion that animal is false can be reached. The process continues visiting the nodes *human* and *lives_in_jungle* before accumulating enough data to conclude that the subject is intelligent.

As shown in the examples given above, MACIE is also able to perform a form of rudimentary explanation of its conclusions, similar to first generation expert systems explanation. Explanations on why a certain conclusion was reached can also be provided by restating the values of the known input concepts in the form of rules. The rules are not explicitly stored in the knowledge base but implicitly derived during inference.

8.3 GENERATING RULES FROM NEURAL NETS

A neural net possesses knowledge which is implicit, local and distributed to the individual neurons and the weight set. This makes the justification and explanation of its actions difficult. This black box feature of neural nets is a major disadvantage when compared with expert systems. Various algorithms have been proposed for extracting a knowledge base of rules from a trained neural net. (Gallant, [2], Sestito and Dillon, [12], Nikolopoulos, [10]). The extracted rule base can be used to append explanation facilities to a neural net. It can also be viewed as an alternate machine learning technique for deriving rules from raw data.

Consider a feedforward neural net with an input layer of n cells, $c_1, \ldots, c_n$, and one output cell c. Let the corresponding weights be $w_1, \ldots, w_n$ and the activation function the discrete step function. In this case, the net is equivalent to a boolean expression in the variables $c=f(c_1, \ldots, c_n)$ expressed as a disjunction of conjunctions. This disjunctive expression can be also expressed as a set of rules with one rule for each conjunction in the clause.

The brute force approach of finding the corresponding rule set leads to combinatorial explosion. We would have to examine all possible rules of the form:

IF c_{i1} and ...and c_{ik} THEN c=1, where $\{i_1, \ldots, i_k\} \subseteq \{1, \ldots, n\}$

for validity and eliminate the rules which are subsumed by other rules (the more general rules have less number of conditions). There are $\sum_k C(n,k)$ such rules, where $C(n,k)=n!/[k!(n-k)!]$ and it takes 2^{n-k} steps in order to verify a rule with k variables in its condition.

One way to ease the combinatorial explosion problem would be to reduce the number of conditions allowed in a rule by allowing conjunctions of at most k ($k \leq n$) conditions. To avoid performing the exponential number of tests needed to verify a rule, the following heuristic can be used ([2]).

Validity test:
A rule is valid if the sum of the absolute values of the weights from each of the k cells is greater than the sum of the absolute values of the rest of the weights.

This simplifies the process of checking for validity into an algorithm of constant order. Consider for example, four input nodes c_1, c_2, c_3, c_4 and one output node c with weights 1,-2,4, and 2 respectively. Let k=2, i.e. search for rules with conjunctive premises of at most two conditions. No rule of one condition passes the test. The following rules pass the validity test for k=2:

IF c2 is false and c3 is true THEN c is true
IF c3 is true and c4 is true THEN c is true

IF c1 is true and c3 is true THEN c is true

And their duals:

IF c2 is true and c3 is false THEN c is false
IF c3 is false and c4 is false THEN c is false
IF c1 is false and c3 is false THEN c is false

The process of eliminating those rules which are subsumed by more general rules is of linear order, but does not apply here as there are no rules for k=1. The derived rule set consists of the most general, valid rules describing the operation of the net.

In the case of multilayered nets the given methodology is not guaranteed to produce all the most general valid rules. Even so, an approximation may be enough. For other heuristics and extensions to algorithms which derive rules from networks of continuous activations, the reader is referred to [2, 10, 12].

8.4 A HYBRID EXPERT SYSTEM FOR INVESTMENT ADVISING

This section presents an example of a hybrid expert system, *Investor*, which combines the connectionist, the symbolic and the evolutionary paradigms in one integrated system. Investor employs a neural network front end which performs classification tasks. The neural network detects the interest rate trends and classifies the user into an appropriate investor model. Investment strategy recommendations largely depend on the interest rates and characteristics of the investor. The investor model and the interest rate information derived by the neural network module are then used by the rule based knowledge base in order to infer the appropriate investment strategy. Genetic algorithms and classifier systems were employed to learn the architecture and train the neural net component of the system.

Factors which contribute to investment strategy recommendations and which are taken into account by the system include the state of the economy and the investor's tolerance for risk, lifestyle, and financial goals. The user interface of Investor asks the user for information about his/her lifestyle, risk tolerance and financial goals. One of the modules of the knowledge base contains a rule set to build a user profile using the acquired characteristics of the user. External factors, such as the state of the economy and interest rate also affect investment buy recommendations.

It is an extremely difficult if not impossible task to construct a rule base for tracking the interest rates, so the neural network paradigm was used to predict if the interest rate is rising, falling or remaining steady. This neural net component of the knowledge base is trained using data on the state of the economy which is listed weekly in many financial publications. The interest rate information derived by the neural net is fed into another rule based module which together with the derived investor profile uses it to derive an appropriate investment strategy.

Investor offers general investment advice. For example, it may recommend to invest in blue chip growth funds without specifically suggesting which ones.The knowledge based component of the system was implemented in PROLOG and the neural network component in C. The system architecture is described below.

8.4.1 System architecture

The basic system modules are the *interface module*, the *neural net*, the *classifier*, the *matcher*, the *plan generator* modules and an *explanation facility*.

The *interface module* determines a user's investment profile, by asking questions about their lifestyle (marriage status, family status, e.t.c.), risk tolerance and investment goals. It also performs error checking, verifies consistency of the input data, outputs explanations generated by the explanation module and retrieves the financial data to train or update the neural net.

The *classifier module* uses the profile derived by the interface module to classify the investor into an appropriate investor model. The investor model then is fed into the *matcher module* to find the most appropriate stored model that matches it. This model is used by the plan generator to derive an appropriate investment plan.

The rules in the classifier module are based on profile attributes of lifestyle, attitude towards risk and investment goals. Lifestyle categories include investors who are recent college graduates, newly married, raising a family, empty nesters and those approaching or who are in retirement.

Using their attitude towards taking risks investors are categorized as conservative, aggressive or venturesome investors. Investment goals can be short, intermediate or long term.

The *plan generator* produces the investment plan based on inputs from the classifier and the Neural Net module. The investment plan generated takes into account the trend of the interest rates and the investor model into which the user has been classified.

The matcher can also analyze available data on the user's present investment practices to determine if these practices are consistent with the strategy recommended by the generated plan. Any currently used investment practices by the investor which are inconsistent with the derived plan are given to the user. Based on the user's profile the explanation module can provide the user with an explanation of why his current practice is not recommendable.

8.4.2 The neural network module

The average maturity of money market funds ranges from 30 to 50 days and is available in Donoghue's Money Market Funds tables. As the average maturity increases, the interest rates are steady or decreasing. Another variable which is used to predict the interest rate trend is the uncompounded yield of taxable money market funds. If the current yield is higher, then it is more likely that interest rates are rising.

The neural network system module is a feedforward neural net with 33 nodes in the input layer, two hidden layers of 3 and 2 nodes respectively and one output node. The threshold function is the linear function (g(h)=h). The first 25 of the input nodes are used to represent the average yield of taxable money market funds for each of the 25 weeks preceding the current week. Seven nodes accept as input the daily yield of the last seven days and the last input node corresponds to the average maturity of taxable money funds. Input vectors were created by randomly picking 100 start points in the time continuum and calculating the average yields for the next 26 weeks from the starting point, with the 26th week considered to be the current week.

The net architecture was configured and the net was trained using the algorithms employed in NEUTOGEN, a neural network configuration system which uses genetic algorithms to train a feedforward net and classifier systems to configure the most appropriate architecture for the net, ([11]). The following section briefly describes the evolutionary learning of neural nets.

8.4.3 Neural network training using genetic algorithms

Genetic Algorithms and classifier systems can be used to determine the most appropriate architecture of a neural net and learn the weight set for the architecture (Miller, [6], Whitley, [13], Nikolopoulos, [7]).

We give below a description of a genetic algorithm which can be used to learn the weights of a neural net of a given architecture. The network links are labeled from 1 to n and each weight value is represented by a binary string of k bits. Individuals in the genetic population consist of n*k bits, with the first k bits representing the weight of link labeled 1, the second substring of k bits representing the weight of link 2, etc.

The Genetic Algorithm is applied in the sequence of steps (see [7]):

(1) Randomly generate an initial population of binary strings where each population individual represents a set of connection weights.

(2) Convert each individual into the corresponding set of weights and feed each pattern of the training data set into the weighted network. The strength of the individual is given by the sum of the squares of the difference between desired output value and actual output value for each training pattern i:

$$\sum_i (D_i - O_i)^2$$

where D_i is the desired output for pattern i and O_i the actual output for the given weight set. The accuracy of a net is inversely proportional to its strength. Rank order the individuals in the population in descending order of strength. The *fitness* of an individual is given by its rank. Larger fitness corresponds to smaller accumulated error and a more accurate neural net.

(3) Apply reproduction, crossover and mutation to produce successive generations of neural nets, until there is an individual which corresponds to a neural net with accumulated error below a

predefined threshold or until a predefined number of generations has been produced.

In addition to algorithms such as the one given above for training neural nets with genetic algorithms, classifier systems can be used in order to configure the appropriate architecture for the net. Neural networks can be converted to equivalent classifier systems and neural network training reduced to classifier system training. For a method which configures the internal structure of the hidden layers and the link weights for the net architecture, the reader is referred to ([11]). These algorithms can automate the ad hoc, trial and error methods currently employed by practitioners during development of neural network applications.

8.4.4 An investment advice session with the expert system

We give below a sample run of the system:

: investment(Advice)?

- 1: Seek Investment Advice
- 2: Update Indicators
- 3: Current Investment Activity

Choice? (1/2/3/e) 2.
(Please wait, activating the neural net)
...

The interest rates are rising.

-1: Seek Investment Advice
-2: Update Indicators
-3: Current Investment Activity

Choice? (1/2/e) 1.

Are you retired or about to be? (y/n/e) n.
Are you raising a family? (y/n/e) n.
Are you newly married? (y/n/e) n.
Are you a recent college graduate? (y/n/e) n.
Have you accumulated cash to cover three months expenses? (y/n/e) y.

Have you set investment goals? (y/n/e) y.

Choose category best describing your willingness to take investment risks.

- 1: Conservative - seek safety, income, and preservation of capital
- 2: Aggressive - willing to take managed risks to achieve gains
- 3: Venturesome - willing to take greater risks searching for more return

Enter choice (1/2/3/e) 3.

Evaluate your goals with respect to the following categories. Choose category best describing the goals you wish to pursue.

- 1: Short-term - one to two years
- 2: Intermediate-term - three to five years
- 3: Long-term - five or more years

Enter choice (1/2/3/e) 3.

Does your company offer a retirement plan? (y/n/e) y.

Advice:

Step 1. Sell any money market mutual funds you are currently invested in.}
(* this is a product of the classifier/matcher module*)
Step 2. Blue chip growth funds are best for long term wealth building.
Step 3. Set up a tax deferred annuity. Use single premium life insurance.
Step 4. You should be invested in 3 to 5 stock market mutual funds.
Step 5. Consider investing in precious metals and even some junk bonds.
Step 6. Contribute the deductible maximum to your employers retirement plan.

Update Investment Signals on a weekly basis.

Save investor profile? (y/n) y
Save investment strategy? (y/n) n

Update Investment Signals on a weekly basis.

8.5 A HYBRID EXPERT SYSTEM FOR INSURANCE POLICY UNDERWRITING

This section describes a rule based system with a hybrid automated knowledge acquisition component. The system combines the evolutionary with the classification tree approach in deriving a knowledge base from raw data in the domain of insurance underwriting (Nikolopoulos, Duvendack, [9]). One of the objectives is to determine when an insurance policy should be cancelled. A knowledge base of rules was directly derived from a set of policy holders' records circumventing the knowledge acquisition bottleneck. One of the goal attributes was the "*terminate policy*" attribute with possible values: *company_terminated*, *policy_holder_terminated* and *active_policy*. There were 19 attributes defining a policy holder's record. The data was a set of vectors of (attribute-value) pairs and contained noise in the form of missing attribute values.

The general evolutionary methodology for learning a rule base was described in Section 6.5.4. In the insurance underwriting system the evolutionary approach taken to generate a knowledge base of rules can be described as follows:

A rule is encoded as a string $a_1a_2...a_{19}g$ where a_i is equal to the value of the ith attribute (encoded as an integer) or * and g is equal to 1, 2 or 3 corresponding to *company_terminated*, *policy_holder_terminated* and *active_policy* respectively. The presence of a * in position i (i=1 to 19) indicates that the ith attribute is not involved in the premises of that rule. The more general a rule is, the more *'s it will contain in its string encoding. The * symbol is given higher probability of selection during mutation in order to ensure generality in the rules. In the initial genetic population, an individual corresponds to a set of rules (the knowledge base) and consists of a concatenation of encoded rules.

Instead of choosing the initial population of knowledge bases randomly, another machine learning technique was used to produce an initial approximation to the knowledge base. The derived knowledge base was then calibrated to create a much more informed initial population for the genetic algorithm.

In the hybrid variation of the expert system, the initial population of knowledge bases was not chosen at random, but it was derived using the set of rules produced by the expert system shell Ginesys (Generic

Inductive Expert System Shell). Ginesys is a shell which emulates a class of inductive learning algorithms and uses redundant rules to reduce classification errors caused by noise in the data. For details on the underlying algorithms, see Gams and Drobinic ([3]). The initial population of individuals (each individual corresponding to a set of rules) to be used by the GA was created from the rule set produced by Ginesys as follows:

Let n be the desired number of rules in the goal knowledge base and $\{r_1, r_2,\ldots,r_k\}$ the set of rules generated by Ginesys using an ID3 type of learning algorithm on the customer records data set. Randomly choose n numbers between 1 and k+1 with k+1 assigned a higher probability of selection than the rest. For each selected i, with $0<=i<=k$, insert the rule r_i into the population individual. Each time i=k+1 was selected a new rule was generated randomly and added to the individual.

Experimental results showed that the hybrid approach of using genetic algorithms to improve rules in a knowledge base generated by another machine learning approach led to increased classification accuracy and quality of the rules produced. For more details on the hybrid insurance underwriting system described here, the reader is referred to [9].

SUMMARY

This chapter presented some possible ways of integrating multiple paradigms into a unified hybrid expert system. Recent expert system application developments follow a trend of combining the symbolic paradigm with the connectionist and evolutionary approaches. The resulting hybrid systems retain the advantages of each paradigm and aim at improving performance over using a single methodology. The model of a neural expert system was presented which implicitly stores a rule base, can contact dialog with the user for accessing information and can provide explanations of its actions eliminating one of the weaknesses of neural nets of being a black box technology. The case studies of two hybrid expert systems in investment advising and insurance underwriting were presented. The investment advising system demonstrates a way in which the different paradigms can be integrated into a single system of a neural net front end and a rule based back end. Finally, the insurance underwriting system is an example of a hybrid expert system which combines the traditional symbolic approach with evolutionary and tree induction machine learning techniques effectively eliminating the knowledge acquisition bottleneck.

REFERENCES

[1] Edwards, J.S., "Expert Systems in Management and Administration", *European Journal Operation Research*, 61, 1992, 114-121.

[2] Gallant, S., "Connectionist Expert Systems", *Communications of the ACM*, 31, no. 2, Feb. 1988, 152-169.

[3] Gams, M., Drobinic, M., "Learning from examples: A uniform view", *International Journal of Man-Machine Studies*, 34, 1991, Academic Press.

[4] Madey,G.R., Weinroth, J., and Shah, V., "Hybrid Systems in Intelligent Manufacturing", In: *Intelligent Manufacturing*, Ed., Dagley, C., Chapman Hall, London, England, 1992.

[5] Medsker, L.R., "The synergism of expert systems and neural network technologies", *Expert Systems Applications*, 2, 1, 1-2, 1991.

[6] Miller, G.E. and Todd, P.M., "Designing Neural Networks using Genetic Algorithms", *Proceedings of the Third Intern. Confer. on Genetic Algorithms*, Morgan Kaufmann, 1989, pp. 379-384.

[7] Nikolopoulos, C. and Hwang, Y.R., "Genetic Algorithms used in training neural networks", In: *Proceedings of the First Golden West Conference on Intelligent Systems*, Reno, NV, 1991, 8-15.

[8] Nikolopoulos, C., Fellrath, P., "A Hybrid Expert System for Investment Advising", *Expert Systems: The International Journal of Knowledge Engineering,*, 11, 4, 1994.

[9] Nikolopoulos, C., Duvendack, S., "A Hybrid Machine Learning System and its Application to Insurance Underwriting", *Proceedings of the First IEEE Conference on Evolutionary Computation, IEEE World Congress on Computational Intelligence*, June, 1994, 692-695.

[10] Nikolopoulos, C., "Deriving Classification Rules from Neural Networks", *Proceedings of the ISIKNH 94, International Symposium on Integrating Knowledge and Neural Heuristics*, May 1994, 141-146.

[11] Nikolopoulos, C., "Evolutionary Topology Configuration of Neural Nets", *Neural Network World, International Journal on Neural and Mass-Parallel Computing*, 5, 1994, 553-565.

[12] Sestito, S., Dillon, T., "Using Single Layered Neural Networks for the extraction of conjunctive rules", *Journal of Applied Intelligence*, 1, 1991, 157-173.

[13] Whitley, D.: The Genitor Algorithm and Selection Pressure. In: *Proceedings of the Third International Conference on Genetic Algorithms*, San Mateo: Morgan Kaufmann, 1989, 116-121.

[14] Zadehi, F., *Intelligent Systems for Business: Expert Systems With Neural Networks*, Wadsworth, 1992.

APPENDIX A

AN INTRODUCTION TO PROLOG

A.1 PROLOG BASICS

In his paper "Algorithms=Logic+Control", Kowalski proposed that the problem specification can be separated from control and declarative logic statements can be interpreted as procedural computer instructions ([9]). These ideas were implemented by Colmerauer and his colleagues of the Artificial Intelligence Group at Marseilles as the declarative logic programming language Prolog, (Programming in Logic), ([5], [6]).

Unlike the traditional *procedural* languages, (Pascal, C, Fortran, etc.), Prolog is a *declarative* programming language. The Prolog programmer expresses in logic syntax the logical specifications about the problem to be solved, ("declares" the available knowledge about the problem domain) and does not need to specify "how" a solution is to be derived. The built in inference mechanism is based on Robinson's resolution principle and contains the necessary control and algorithmic knowledge on how to execute the specifications and retrieve the answers to queries automatically. The search strategy employed is based on Kowalski's linear resolution strategy and set of support. Negation is handled by employing the closed world assumption (see Chapter 2).

Although Prolog may not be the best choice for certain application domains, for example systems programming, it is ideal for knowledge representation and reasoning tasks. Its knowledge representational ability and automated deductive power render Prolog suitable for solving problems in such varied domains as expert systems, database systems, natural language processing and computer aided design, (Subrahmanyan, [14]).

The basic unit of Prolog programming is a Horn clause, i.e., a clause with at most one positive literal. Such clauses are equivalent to inference rules with only one literal at the rule head. The Prolog rule syntax is of the form:

p:-$q_1, q_2, \ldots, q_n$.

where p, q_i, i=1, . . . ,n are literals.

The above formula is interpreted as:

If $(q_1$ *and* q_2 *and ... and* $q_n)$ *then p.*

In Prolog, a comma is used to indicate the logical AND. Implication is written from right to left and denoted by ":-". A period is used to indicate the end of a prolog statement or query. The left side of the rule is called the *head* of the rule and can only consist of one literal. The right side of the rule is called its *body*. Literals contained in the body of a rule can also be joined by the logical OR operator which in prolog is indicated by a semicolon ";". A rule with a disjunctive body can be rewritten without disjunctions. For example, the rule

c:-a;b.

is equivalent to the set of two rules:

c:-a.
c:-b.

A set of rules, which have the same head predicate with the same arity, is called a *procedure*. A Prolog predicate and its arguments are represented by *predicate_name(arguments)*. The arguments may be variables, constants or functions. A variable starts with an uppercase character while predicates, constants or functions with lowercase.

Here is an example of a rule:

color(X,C):-part_of(X,Y),color(Y,C).

This rule could be used to indicate that:

IF(object X is part of an object Y and the color of Y is C) THEN the color of X is C.

The variables in the head of the rule are universally quantified, while the ones that only appear in the body are existentially quantified. For example, the rule given above corresponds to the logic formula:

$$\forall X \forall C[(\exists Y(part\_of(X,Y) \wedge color(Y,C))) \Rightarrow color(X,C)]$$

A fact is a rule whose body is always true. For example, the following fact indicates that Smith is a faculty member in the College of Engineering:

faculty(smith,engineering).

The user interacts with a Prolog program by typing in expressions to be evaluated (queries or goals). Evaluation entails proof of the goals by the automated inference engine. After a successful proof, the bindings made to query variables are given as the answer to the user query, or in the absence of variables in the query the answer *yes* is provided (see Green's method in Chapter 2). In case the proof fails, the answer *no* is given to indicate that the query is unsatisfiable and therefore false according to the CWA.

The Prolog knowledge base given below will be used in the sequel for explaining basic Prolog programming concepts. Figure 3.1 displays the knowledge base in the form of a semantic network. A semantic network is a knowledge representation technique where knowledge is displayed as a directed labeled graph. The nodes correspond to concepts or objects and the arcs correspond to relations between the concept nodes. The arc labels indicate the relation names. An arc (relation) can be bidirectional, like *"is the brother of"*, or unidirectional like *"is a faculty in"*. Knowledge representation techniques including semantic networks will be examined in more detail in Chapter 5.

The knowledge expressed in the semantic network of Figure A.1 can be expressed in Prolog syntax as follows.

location(engineering,bradleyhall).	*(F1)*
location(business,bradleyhall).	*(F2)*
dept_in(electricalengineering, engineering).	*(F3)*
dept_in(civilengineering, engineering).	*(F4)*
dept_in(finance, business).	*(F5)*
status(engineering, accredited).	*(F6)*
is_faculty_in(smith, electricalengineering).	*(F7)*
is_faculty_in(jones, civilengineering).	*(F8)*
is_faculty_in(X,Y) :- dept_in(Z,Y), is_faculty_in(X,Z).	*(R1)*
location(X,Y) :- dept_in(X,Z), location(Z,Y).	*(R2)*
status(X,Z) :- dept_in(X,Y), status(Y,Z).	*(R3)*

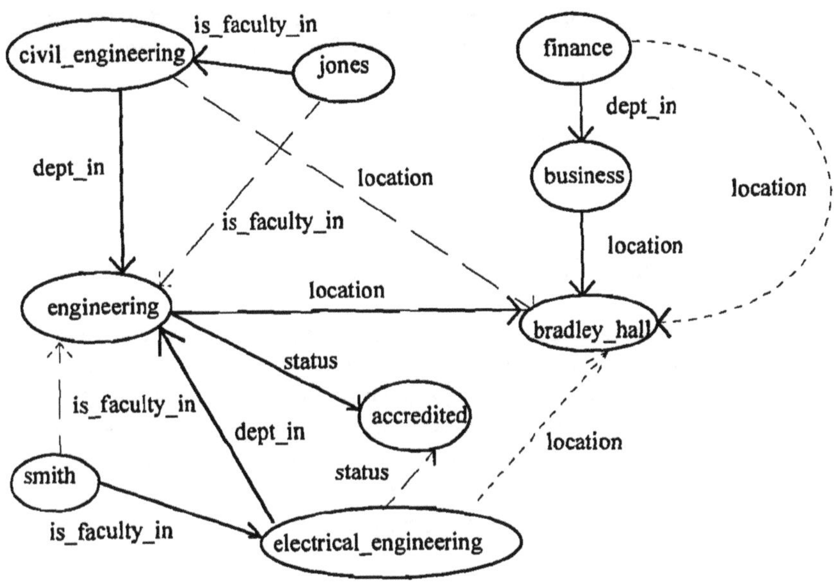

Figure A.1 A semantic net representation of the college knowledge base.

Observe that in the college knowledge base, the facts precede the rules. In general, the prolog programmer has to insert facts before rules in the knowledge base. As it will be explained later, a search for a proof is done in a sequential, depth first fashion with backtracking. Facts can be the stopping criteria for recursion, so facts out of place can cause infinite loops during execution of recursive prolog rules.

Rules (R1), (R2) and (R3) are *tail recursive* rules. They are preferable to *head recursive* rules which can lead to infinite loops. For example, if rule (R2) is rewritten as:

location(X,Y):-location(Z,Y),dept_in(X,Z).,

then a query such as

?location(engineering,sisson).

will cause an infinite loop.

A reader with background in database management systems may

recognize the similarities of a Prolog knowledge base and a relational database. A collection of facts under the same predicate corresponds to a relational table and rules correspond to relational views. In contrast to relational databases, which lack recursive abilities, a Prolog knowledge base allows recursion. Prolog can easily form the basis for implementing a deductive, object oriented database system, ([11]).

A.2 THE PROLOG INFERENCE ENGINE

A Prolog program is a collection of rules and facts, which form the knowledge base. It expresses the programmer's knowledge for the particular problem domain in a declarative manner and is not merely a set of procedural instructions addressed to the computer. Since Prolog inference is based on backward chaining, the knowledge base is utilized by "querying" it. Using a text editor the knowledge base can be created and then loaded into the Prolog working memory, usually by a command such as

consult('filename').

At the Prolog prompt, queries on the consulted knowledge base can then be issued. The built in inference engine of Prolog uses a depth first search strategy with backtracking based on the set of support resolution of Horn clauses, (see Chapter 2). A few examples are given below, on how the inference engine processes queries.

Consider the college knowledge base of (F1) to (R3). The query "Is civil engineering one of the departments in engineering?" can be formulated in Prolog as:

?dept_in(civil_engineering, engineering).
yes

The Prolog inference engine works in top down fashion. It unifies the query with fact (F4), so the answer to the query is *yes*.

Next consider the query "Which college does the department of electrical engineering belong to?", which can be formulated in Prolog format as:

?dept_in(electrical_engineering, L).

During unification the query is unified with fact (F1) by *binding* the variable L to the constant *bradleyhall*. The values bound to query variables are displayed in response to the query after a proof is complete, as follows:

?dept_in(electrical_engineering, L).
L=engineering

Both of the queries given above were deterministic, in the sense that there is exactly one proof path and one answer to them. We give below an example of a nondeterministic query for which there is more than one answer. Some Prolog implementations (for example sb-prolog) print the first answer to the query and then wait for the user to ask for alternate answers, while others (for example Prolog-86) find and print all the alternate answers.

Consider the more complicated query "find all the departments which belong to engineering". In Prolog format the query is expressed as follows:

?dept_in(D, engineering).

After the first match and unification with fact (F3) the variable D is bound to the constant *electrical_engineering*. This is the first answer to the query. A semicolon ";" followed by the <return> key, after each binding is exhibited causes the inference engine to continue searching for possible alternate bindings which also provide an answer to a query. For example, typing ";" after the first answer *D=electrical_engineering*, causes the last binding for D to be undone and the search is continued forward from the position of the last binding. The query *dept_in(D,engineering)* is then successfully unified with fact (F4) under the binding *D=civil_engineering*. Another semicolon will not produce further answers, as the continued search will produce no more matches. So, all the departments, which belong to engineering, are produced by repeated use of the *or* operator(semicolon) after each alternate answer, until the interpreter returns a "no" (meaning no more matches).

?dept_in(D,engineering).
D=electrical_engineering;
D=civil_engineering;
no

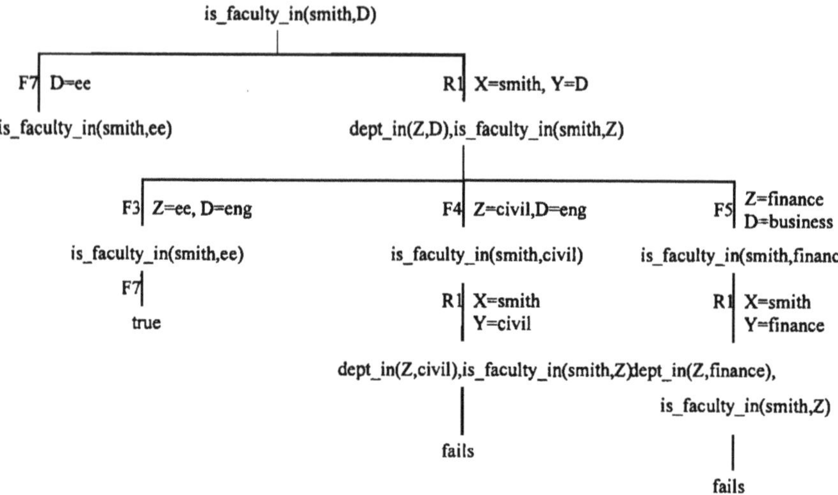

Figure A.2 AND/OR proof tree for query ?is_faculty_in(smith,D).

As a third example, consider the query "which department or college is Smith a faculty in?", written in Prolog as:

?is_faculty_in(smith,D).

The query answering process can be depicted by the AND/OR search tree shown in Figure A.2. The root of the tree is the original query. The nodes of the tree are conjunctions of subgoals that are to be proven during backward chaining. A subgoal that matches the head of more than one rule, becomes an OR node with one outgoing branch for each alternate way of pursuing the proof. When a path down the proof tree leads to an unsatisfiable node, the search engine backtracks to its first ancestor OR node, which has unexplored search branches, undoes the bindings that led to the unsatisfiable node, and pursues an alternate branch.

The given query unifies with fact (F7) under the binding:

D=electrical_engineering.

Typing a semicolon after the binding is exhibited by the system, continues the search of the AND/OR tree for possible alternate bindings that validate the query. The last binding for D is undone and search continues on from the point where the last binding occurred, i.e. Fact (F7). The query fails to unify with any more facts, so the rule (R1) is used. The query can be unified with the rule head by binding X to *smith* and Y to D. Under this binding the rule head will be true if both of the literals in the conjunction at the rule tail are true. So, search proceeds backwards trying to verify the subgoals :

dept_in(Z,D) and *is_faculty_in(smith,Z).*

The first conjunct matches fact (F3) by binding Z to electrical engineering and D to engineering. The second subgoal thus becomes

is_faculty_in(smith,D)

and the search for proof of this subgoal starts again from the top of the knowledge base, matching with fact (F7). Since both of the conjunct subgoals were verified as true the original goal is true and the value bound to D is returned as an answer to the query.

?is_faculty_in(smith,D).
D=electrical_engineering;
D=engineering;
no

Typing a ";" undoes the binding unifying (F5) with the subgoal *dept_in(Z,D)*. The search continues from (F3). Using (F4), Z is bound to civil_engineering and D to engineering. The second subgoal in the conjunctive proof tree node becomes:

is_faculty_in(smith, civil_engineering).

In order to verify this subgoal, the search for a proof starts at the top of the knowledge base. The subgoal can be unified with the head of rule (R1) to yield the conjunctive subgoal:

dept_in(Z,civil_engineering), is_faculty_in(smith,Z).

The subgoal *dept_in(Z,civil_engineering)* fails, and as a result *is_faculty_in(smith, civil_engineering)* also fails. Backtracking to *dept_in(Z,D)* takes place. The next branch of the OR node is followed and fact (F5) is used to bind Z to finance and D to business. The subgoal *is_faculty_in(smith,finance)* is pursued by applying rule (R1) and fails. Since all disjunctive branches of *is_faculty_in(smith,D)* have been tried, the search for alternate proofs terminates.

A.3 NEGATION

Prolog employs the *closed world assumption* (CWA) hypothesis in order to deal with negation, (see Chapter 2, Section 10). Everything that cannot be deduced from the knowledge base is assumed to be false (negation by failure). The built in predicate *not* takes one argument, which is a conjunction of literals and returns true if the search for a proof for the argument fails, false otherwise.

Consider for example the query:

?not(dept_in(electrical_engineering, engineering)).
no

The answer to the query is false, because *dept_in(electricalengineering, engineering)* is proven true.

On the other hand the query:

?not(dept_in(electrical_engineering,business).
yes

was found to be true, since no proof can be reached for the query *dept_in(electricalengineering, business)*.

The query

?not(dept_in(electrical_engineering, X)).
no

is intended to answer the question "Is it true that electrical engineering is not a department in any division?" and not "which of the colleges does the dept. of electrical engineering not belong to?". A variable cannot be assigned a binding inside the *not* predicate. The answer to the query is false, since there is a binding to satisfy the query: *dept_in(electrical_engineering,X)*.

The query

?dept_in(finance,X), not(dept_in(civil_engineering,X)).
yes

returns true since X is bound to business when the *not* subgoal is reached.

If the negation of a predicate is part of a conjunction, care has to be taken, because a variable cannot be bound inside the *not* predicate. For example, in order to find all colleges to which electrical engineering does not belong and which are located in bradley hall, the following query will not work:

?not(dept_in(electrical_engineering,D)),location(D,bradleyhall).
no

The answer to the query is *no,* since the subgoal *dept_in(electrical_engineering, D)* succeeds for D=engineering and therefore *not(dept_in(electrical_engineering,D))* is false.

Instead the query:

?location(D,bradleyhall),not(dept_in(electrical_engineering,D)).
D=business

should be used.

A.4 ARITHMETIC

Although Prolog is not designed for number crunching applications and should not be used for such, it does have the ability to perform numeric calculations. The notation used is the infix notation. There are a number of built in relational operators:
< (less than), > (greater than), <=, >=, = (equal), <> (not equal).

Comments in Prolog code are preceded by the % symbol.

For example:
% a rule to classify an individual as middle class, if his/her income is
% between 30000 and 60000.
income(smith,58000).

```
middle_class(X):- income(X,Y), Y>=30000, Y=<60000.

?middle_class(smith).
yes
```

A.4.1 The assignment operator

The assignment operator is denoted by "is", while "=" is used as the test for equality.

Examples:

```
% Find the sum of X and Y and store it in Z.
sum(X,Y,Z):- Z is X+Y.

?sum(3,4,Z).
Z=7
?sum(3,1,5).
no
?sum(3,4,7).
yes
% arithmetic procedures are usually not reversible
?sum(3,Y,7).
no
```

If the right side of the assignment operator contains an unbound variable, the result is failure. If the left and right side are both bound, then if they are equal the predicate succeeds, else it fails. A useful extra-logical predicate is the built in *var(X)* predicate. It has one argument and returns true if the argument is a free variable, and false if the argument is bound.

For example, we could expand the usefulness of the sum procedure by adding the rule:

```
sum(X,Y,Z):-not(var(X)),var(Y),not(var(Z)), Y is Z-X.
?sum(3,Y,7).
Y=4
```

This rule enables the success of the predicate "sum" in the case where Y is a variable and X, X are bound. It evaluates Y to the difference of Z and X.

A.4.2 The comparison operator

The comparison operator returns the value true if the expressions on its left and right side are unifiable. Any unbound variables on either side become bound to the substitution required for unification.

Examples:

equal(X,Y):-X=Y.
?equal(3,5).
no

?equal(3,Y).
Y=3;
yes

?equal(3,3).
yes

?equal(ab,ab).
yes

?equal(X,rich(mike)).
X=rich(mike)

?equal(father(X,mike),father(joe,Y).
X=joe, Y=mike

A.4.3 Recursive arithmetic

We give some more examples of using recursion when evaluating arithmetic expressions.

Consider a prolog predicate *fact* of two arguments N and K which evaluates K as the factorial of N, N!=1*2*...*N. For example:

?fact(4,N).
N=24

The product 1*2*...*N is equal to (1*2*...*(N-1))*N, so the factorial of N is equal to N times the factorial of N-1. To avoid an infinite loop in evaluating the recursion we need a stopping criterion: the factorial of 0 is 1. In Prolog syntax, the procedure *fact* is coded as follows:

```
fact(0,1).
fact(N,M):-K is N-1, fact(K,X), M is N*X.
```

Observe that *fact* is not a reversible predicate. The query *?fact(N,24)* will not return N=4, but it will lead to an error. This is usually the case with arithmetic predicates, since the assignment operator "is" doesn't allow unbound variables in its assignment side.

As a final example, consider the following program which evaluates X^N. The predicate *exp* accepts three arguments X,N and Z and evaluates Z as X to the power N. The recursive rule states that $X^N=(X^{N-1})*X$. The stopping criteria state that *1 to any exponent is equal to 1* and that *anything raised to the exponent 0 is equal to 1*. The underscore "_" is used in Prolog to indicate an *unnamed variable* (a variable that does not need to pass its binding to other subgoals can be indicated by underscore instead of a variable name).

```
exp(1,_,1).
exp(_,0,1).
exp(X,N,Z):-M is N-1, exp(X,M,R), Z is R*X.
```

A.5 LISTS

Lists are a basic tool for implementing more complex data structures in Prolog and are extensively used in many applications including text processing. A list is an atomic element in the sense that variables can be bound to lists. A list is represented in Prolog by a sequence of elements enclosed in brackets and separated by commas.

For example, *[saturday,swimming,sunday,church]* is a list of four elements. Lists can be embedded in other lists. For example,

[[saturday,swimming],[sunday,church]]

is a list of two elements, each of which happens to be a sublist. The empty list is represented by [ ].

Given the fact:

schedule([saturday,swimming,sunday,church]).

the answer to the query

?schedule(X).

is *X=[saturday,swimming,sunday,church]*

while the answer to the query

?schedule(X,Y,Z,W).

is *no* (failure), since the predicate *schedule* is a predicate of one argument (which happens to be a list), while in the query the predicate schedule has four arguments making unification impossible.

In contrast to the query given above, the query

?schedule([X,Y,Z,W]).

can be matched with the fact involving the predicate *schedule* and gives after unification:

X=saturday, Y=swimming, Z=sunday, W=church

The first element of a nonempty list is called the *head* of the list, and the remaining part of the list is called the *tail* of the list. The head is an atom or a list, while the tail of a list is always a list itself.

For example, saturday is the head of the list *[saturday, swimming, sunday, church]* and *[swimming, sunday, church]* is the tail. As another example, consider the list *[[a,b],c]*. Its head is the list *[a,b]* and its tail the list *[c]*.

The built in functor "|" is used to extract elements from a list, using the syntax [A1,A2,...,An|L]. The left side of the functor can include one or more arguments (constants or variables) which are atoms and the right side one argument which is a list. During unification with a list M, A1,A2,...,An are matched to the first n elements of the list M and L is bound to the remainder of the list M. For example, given the fact *schedule([saturday,swimming,sunday,church]).*

the query

?schedule([X|L]).

will produce

X=saturday, L=[swimming,sunday,church]

Another example:

?schedule([X,Y|L]).
X=saturday, Y=swimming, L=[sunday,church]

As a final example consider the query *?schedule([X,Y,Z,W,G|L]).* This query fails, since the list in the *schedule* predicate has only four arguments and unification to the five atoms of the query cannot succeed.

The reader who is familiar with the list processing language LISP will recognize that [A|L] is the equivalent of the LISP *car* function.

A.5.1 Some basic list processing functions

We give some additional examples of list processing procedures written in Prolog.

1. The equivalent of the LISP function *car* which extracts the head of a list can be written in Prolog as:

car([X|L],X).
or as:
car([X|L],First):-First=X.

Given the query

?car([3,4],X).

the answer is X=3, since during pattern matching X is bound to 3 and L to the list [4].

Similarly, the LISP "cdr" function, which gives the remainder of a list after its head is deleted, can be written in Prolog as:

```
cdr([X|L], L).
```

which when queried by

```
?cdr([3,4,5],L).
```

gives *L=[4,5]*

2. A function to determine if an atom belongs to a list:

```
member(X,[X|L]).
member(X,[Y|L]) :- member(X,L).
```

This is a recursive predicate. The first rule is the stopping criterion of the recursion. If X is bound to a value equal to the head of the list, then X is a member of the list. If X is bound to a value not equal to the head of the list, then X is a member of the list if and only if X is a member of the tail of the list.
When queried the *member* predicate gives:

```
?member(3,[4,c,5]).
no

?member(3,[4,3,5]).
yes

?member(X,[4,3,5]).
X=4;
X=3;
X=5;
no
```

3. Function to find the length of a list (number of atoms in the list). The length of the empty list is 0. The length of a nonempty list is equal to the length of its tail plus 1.

```
length([ ],0).
length([X|L],N):-length(L,M), N is M+1.
```

When queried it gives:

```
?length([1,2,3], N).
N=3
```

4. Concatenate two lists.
The predicate *append(L,M,N)* concatenates lists L and M to produce list N. The result of appending the empty list [ ] to a list L is L. The result of appending a list [X|L], whose first element is X, to a list M, is a list whose first element is X and whose tail is the appendum of L to M. This is expressed in Prolog as follows:

```
append([], L, L).
append([X|L],M,[X|N]):-append(L,M,N).
```

5. Delete an element from a list.
The predicate *delete(X,L,M)* of three arguments (an atom X, and two lists L and M) deletes all occurrences of atom X from the list L to produce the list M. When an element is deleted from the empty list, the result is the empty list. To delete all occurrences of an element X from a list whose head is equal to X, delete the head of the list and delete all occurrences of X from the tail of the list.

```
delete(X,[],[]).
delete(X,[X|L],M):-delete(X,L,M).
delete(X,[Y|L],[Y|M]):-not(X=Y),delete(X,L,M).
```

When queried gives:

```
?delete(a,[a,b,a,c],M).
M=[b,c];
M=[b,a,c];
M=[a,b,a,c];
no
```

6. To reverse a list:
The predicate *reverse* of two arguments returns in the second argument the reverse list of the list bound to the first argument. The reverse of the empty list is the empty list. The reverse of a list L, whose first element is H and the tail T, is the reverse of the tail T of L appended to the list

[H].

```
reverse([],[]).
reverse([H|T],R):-reverse(T,RT),append(RT,[H],R).
```

When queried gives:

```
?reverse([a,b,c],M).
M=[c,b,a]
```

7. Union of two lists:
Union is a predicate of arity 3, union(M,N,L), which accepts two lists M and N and calculates their union L.

For example,

```
?union([2,3,4],[3,5,4,6],L).
L=[2,3,5,4,6]
```

IF the first element of list M is a member of the list N, THEN the union of lists M and N is equal to the union of the tail of M and N, ELSE the union of M and N is equal to the first element of M followed by the union of the tail of M and N.

```
union([],List,List).
union([Head|Tail],Other_list,New_list):-
                                member(Head,Other_list),!,
                                union(Tail,Other_list,New_list).
union([Head|Tail],Other_list,[Head|New_list]):-
                                union(Tail,Other_list,New_list).
```

8. Eliminate duplicate elements from a list:
The *eliminate* predicate accepts two lists M and L as arguments, and bounds the second list to the first after eliminating the duplicate elements of the first list. For example:

```
?eliminate([2,3,2,4,3],L).
L=[2,4,3]
```

In order to eliminate duplicates from the list M yielding the list L:
IF the head element of list M is a member of the tail of the list M, THEN the list L is equal to the list we get by eliminating duplicates from the tail

of M, ELSE the list L is equal to the list whose head is the head of M and whose tail is the list we get by eliminating duplicates from the tail of M.

```
eliminate([ ],[ ]).
eliminate([X|L],M):-member(X,L),eliminate(L,M).
eliminate([X|L],[X|M]):-eliminate(L,M).
```

9. Check if a list is a prefix of another list:
The *prefix* predicate accepts as input two lists and returns true if the first list is a prefix of the second. For example,

```
?prefix([2,4],[2,4,5,6]).
```

returns true, while

```
?prefix([2,5],[2,4,5,6]).
```

returns false.
The empty list is a prefix to any other list. If the list is not empty, then List1 is a prefix to List2 iff they have the same head and the tail of List1 is a prefix of the tail of List2.

```
prefix([],_).
prefix([Head|Tail1],[Head|Tail2]):-prefix(Tail1,Tail2).
```

10. Determine if a list M is a sublist of a list N.
A list M is a *sublist* of N iff the sequence of elements in M is a contiguous subsequence of N. IF lists M and N have the same head and the tail of M is a prefix of the tail of list N, THEN list M is a sublist of N, ELSE if list M is a sublist of the tail of N then list M is a sublist of N.

```
sublist([Head|Tail1],[Head|Tail2]):-prefix(Tail1,Tail2).
sublist(List1,[_|Tail2]):-sublist(List1,Tail2).
```

```
?sublist([3,4],[3,5,3,4]).
yes
?sublist([2,4],[3,2,5,4]).
no
```

A.6 EXTRA-LOGICAL PREDICATES

A number of extra-logical predicates have been introduced in Prolog in order to improve efficiency.

A.6.1 The cut

The *cut*, denoted by !, is a predicate of no arguments and allows the programmer to control the search for a proof by pruning branches from the proof tree. The predicate at the head of the rule with the cut, is considered to have failed when the cut is encountered during backtracking. The cut is always satisfied on forward subgoal expansion, but on backtracking it always fails. Furthermore, the goal that called the

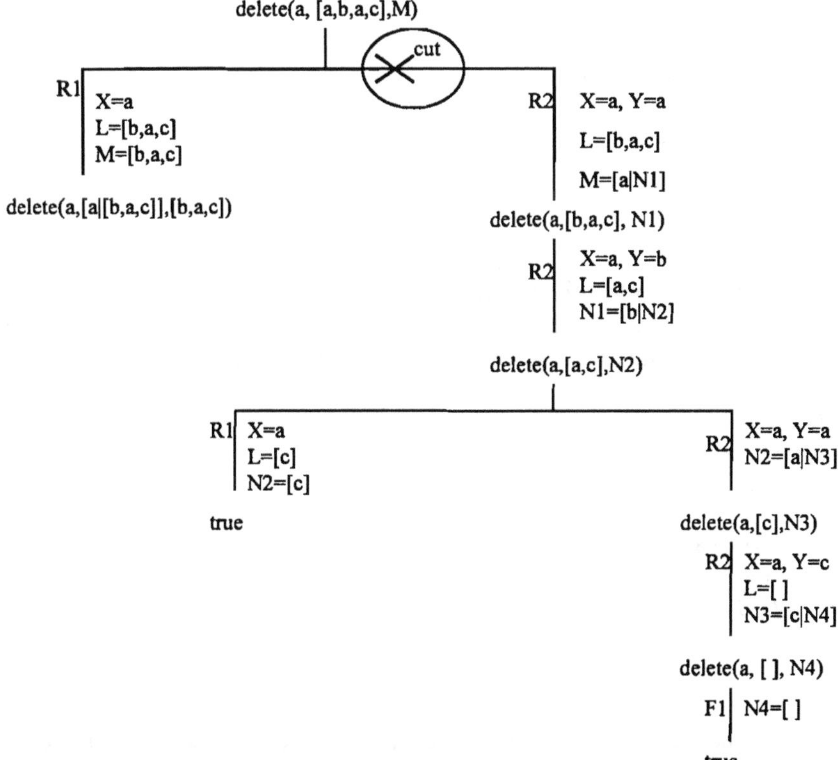

Figure A.3 The AND/OR proof tree for ?delete(a,[a,b,a,c],M).

rule containing the cut fails. All other rules following the rule with the cut and which share the same head are aborted as well.

To demonstrate the effect of the cut consider the following examples.

Suppose we want to write a version of the *delete* predicate, *delete(X,L,M)*, which deletes only the first occurrence of X in the list L and binds the result to the list M. The cut can be used to achieve this as follows:

```
delete(X,[],[]).
delete(X,[X|L],L):-!.
delete(X,[Y|L],[Y|M]):-delete(X,L,M).
```

The query

```
?delete(a,[d,a,b,a,c],M).
```

produces the answer

```
M=[d,b,a,c];
no
```

The search terminates after the first deletion due to the cut which kills the *delete* goal. If the cut is omitted from the second rule, then more than one occurrence of an element can be deleted as:

```
delete(X,[],[]).
delete(X,[X|L],L).
delete(X,[Y|L],[Y|M]):-delete(X,L,M).

?delete(a,[a,b,a,c],M).
M=[b,a,c];
M=[a,b,c];
M=[a,b,a,c];
no
```

The effect of the cut on the proof tree of *predicate delete* is shown in Figure A.3.

Another example of how the *cut* affects a program is given below.

Given that *cdc* and *cat* are the names of two local companies. Consider the following knowledge base, where the facts and rules have been labeled for future reference:

```
(F1)  local(cdc).
(F2)  local(cat).
(F3)  growing(cdc).
(F4)  growing(cat).
(F5)  nolayoffs(cdc).
(F6)  stocksup(cat).
(R1)  invest_in(X):-local(X),stable(X).
(R2)  stable(X):-growing(X),stocksup(X).
(R3)  stable(X):-nolayoffs(X).
```

When queried on which company to invest in:

?invest_in(X).

the answer is:
X=cdc;
X=cat;
no

If the cut is introduced in the second rule and rule (R2) is replaced by (R2')

```
(R2)  stable(X):-growing(X),!,stocksup(X).
```

then the query *invest_in* gives:

?invest_in(X).
X=cat;
no

The proof tree is shown in Figure A.4. During traversal of the proof tree for the program with the cut, the disjunctive subgoal *stable(cdc)* is reached.

Following the branch with rule (R2) leads to a dead end. During backtracking, the cut is encountered and its effect is to kill the subgoal *stable(cdc)* and the other disjunctive branches are not tried. So backtracking goes all the way to the conjunctive subgoal *local(X).*

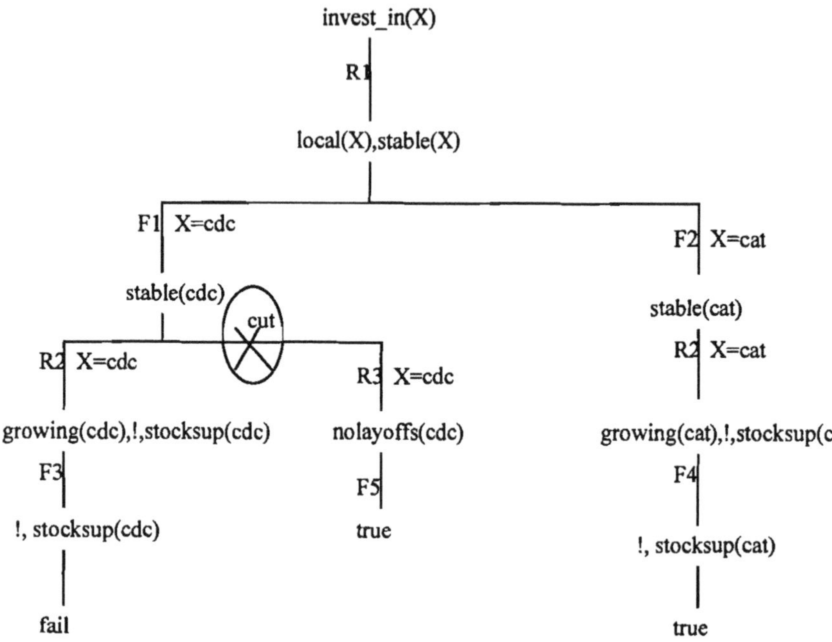

Figure A.4 The AND/OR proof tree for ?invest_in(X).

A.6.2 Other extra-logical predicates

A.6.2.1 Built in predicates

An additional list of useful extra-logical predicates is given below.

var(X) Is true iff X is an unbound variable, false otherwise.

see(Filename) Opens a file for input.

read(X) Reads from the currently open input file and bounds X to the input value. If no file has been opened, then input is read from the screen.

seen Closes the input file opened by *see* and the screen is reset as the input medium.

tell(Filename) Opens file Filename as output file.

write(X) Writes the value bound to X into the file currently open by *tell* or if no output file has been opened, it writes on the screen.

told Closes the current output file and resets the screen as output medium.

fail Forces backtracking as it is always false.

clause(A,B) Creates a rule with head equal to A and body B, A:-B.

S=.. L Where S is a term and L is a list. A list is L is converted into a term S and vice versa.

findall(Term,Query,List)
Finds all the answers to the query *Query* and puts them in *List* in the format described by *Term*. (In some versions of Prolog this predicate is called *bagof* instead of *findall*)

An example of how *findall* works is given below. Given the facts:

status(engineering,accredited).
status(computerscience,accredited).

the query *?findall(X,status(X,accredited),L).*

returns the answer: *L=[engineering,computerscience]*

and the query *?findall(accredited(X),status(X,accredited),L).*
returns: *L=[accredited(engineering),accredited(computerscience)]*

We give below an example of how the universal predicate "=.." works.

?location(computerscience,bradleyhall)=..L.
L=[location,computerscience,bradleyhall]

?S=..[location,computerscience,bradleyhall].
S=location(computerscience,bradleyhall)

The prolog predicate *call* issues its argument as a query to the system. For example:

?call(location(ee,X)).

returns: *X=bradleyhall.*

As an example of using the *call* and =.. predicates, consider the following Prolog procedure to implement the Lisp function *mapcar*. *mapcar* applies a given operator to each individual element of a list to derive a new list. For example, suppose that *add1* is defined as: *add1(X,Y):-Y is X+1.* Then

```
?mapcar(add1,[1,4,5],L).
L=[2,5,6]
```

Procedure *mapcar* can be defined as follows:

```
mapcar(Op,[ ],[ ]).
mapcar(Op,[H1|T1],[H2|T2]):-
        Goal=..[Op,H1,H2],call(Goal),mapcar(Op,T1,T2).
```

A.6.2.2 Updating the knowledge base

The following predicates can alter the knowledge base by adding or deleting knowledge. If the variable *Fact* is bound to a Prolog fact or rule, then:

asserta(Fact) Adds the value bound to Fact as a fact in front of all the facts and rules with the same predicate name.

assertz(Fact) Appends the value bound to Fact as a new fact in the knowledge base after all the facts and rules with the same predicate name.

retract(Fact) Deletes the fact equal to the value bound to Fact from the knowledge base.

A.7 A PROLOG QUESTION ANSWERING SYSTEM

The following example shows how a decision tree knowledge structure can be implemented in Prolog.

An investor is to be advised which stock to buy into: oil, computers or telecommunications. The advice is based on the type of portfolio the investor is to maintain. For example, investors can be classified into one of the following categories, based on the type of portfolio:

high_risk, moderate_risk, or *stable_risk* portfolios.

A sample classification tree incorporating the knowledge used for determining the portfolio type is shown in Figure A.5.

A decision on which stock to invest in can be made based on the investor's portfolio type. An investor suited for a stable risk portfolio should invest in oil, a moderate risk portfolio in telecommunications and a high risk in computers.

The rule base representation of the investment knowledge given above can then be expressed in Prolog as follows:

```
moderate_risk(X):-ask_marital_status(X,Y), Y=married,
        ask_income(X,I), I<=50000, ask_mortgage(X,Z),
        Z<=50000,!.
moderate_risk(X):-
        ask_marital_status(X,M),M=married),ask_income(X,I),
        I>50000,!.
moderate_risk(X):-ask_marital_status(X,M), M=single,
        ask_income(X,I), I<=35000,!.
stable_risk(X):-ask_marital_status(X,M), M=married, ask_income(X,I),
        I<=50000, ask_mortgage(X,Z), Z>50000,!.
stable_risk(X):-ask_marital_status(X,M), M=single, ask_income(X,I),
        I>35000, ask_age(X,A),A>50,!.
high_risk(X):-ask_marital_status(X,M),M=single, ask_income(X,I),
        I>35000, ask_age(X,A),A<=50,!.
invest(X,oil):-stable_risk(X),!.
invest(X,telecommunications):-moderate_risk(X),!.
invest(X,computers):-high_risk(X),!.
```

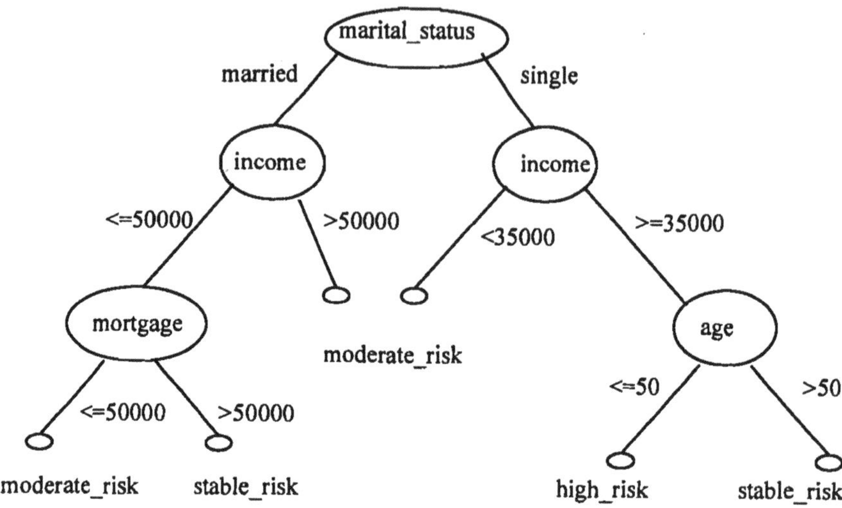

Figure A.5 Decision tree for portfolio type classification.

We now build a input/output interface module to extract the needed information about income, mortgage, age and marital status from the user.

```
main(X,Z):-var(X), write('what is your name?'),read(X), invest(X,Z),!.
main(X,Z):-invest(X,Z),!.
ask_marital_status(X,Y):-marital_status(X,Y).
ask_marital_status(X,Y):-
        not(marital_status(X,Y)), write('what's your marital status:
        married or single?'), read(Y), nl,
        asserta(marital_status(X,Y)).
ask_income(X,Y):-income(X,Y).
ask_income(X,Y):-
        not(income(X,Y)),write('wha's your annual
        income?'),nl,read(Y), asserta(income(X,Y)).
ask_mortgage(X,Z):-mortgage(X,Z).
ask_mortgage(X,Z):-
        not(mortgage(X,Z)),write('what is your
        mortgage?'),read(Z),nl, asserta(mortgage(X,Z)).
ask_age(X,A):-not(age(X,A).
ask_age(X,A):-
        not(age(X,A)), write('what is your age?'), read(A),
```

```
        nl,asserta(age(X,A)).
```

Here is a sample run of the program:

```
?main(X,Z).
what is your name?
smith
what is your marital status?
married
what is your income?
45000
what is your mortgage?
60000
X=smith, Y=oil
```

Questions about income, marital status, etc. are not repeated after they are asked once. This is achieved as follows. Every time an answer to a question is supplied by the user, the answer is appended to the knowledge base as a fact. Before asking a question we check to see if there is a fact already appended in the knowledge base providing the answer, in which case the question is not repeated.

A.8 PROCEDURAL PROGRAMS IN PROLOG

Even though Prolog was not intended as a procedural language and should not be used as such, we show in this section that it is possible to implement procedural programs in Prolog.

A.8.1 If-then-else

Consider the IF control structure: If X=0 Then A=1, Else A=2;

It can implemented in Prolog as the predicate *if(X,A)*:

```
if(X,A):-X=0,A is 1.
if(X,A):-not(X=0), A is 2.
```

or using the cut:

```
if(X,A):-X=0,!,A is 1.
```

if(X,A):-A is 2.

A.8.2 Linked lists

The linked list of Figure A.6 can be implemented in Prolog as a set of facts of the form:

list(Previous_pointer, Data, Next_pointer),

where the arguments correspond to the list node identifier, the data stored in the node and the pointer to the next node of the linked list. The predicate *head(List_name, Head_pointer_id)* is also used to indicate the identifier number of the first element of the list *List_name.*

Head(names_list,5).
list(5,basil,10).
list(10,chris,3).
list(3,christian,0).

A predicate *delete(P)* to delete the element at position P of the list *List_name* is given below:

```
delete(P, List_name):- head(List_name, Head), Head=P, list(P,D,N),
                        retract(head(Head, P)), assert(head(Head, N)).
delete(P,List_name):-list(P,D,N),list(A,B,P),L is N, retract(list(P,D,N)),
                        retract(list(A,B,P)), asserta(list(A,B,L)).
```

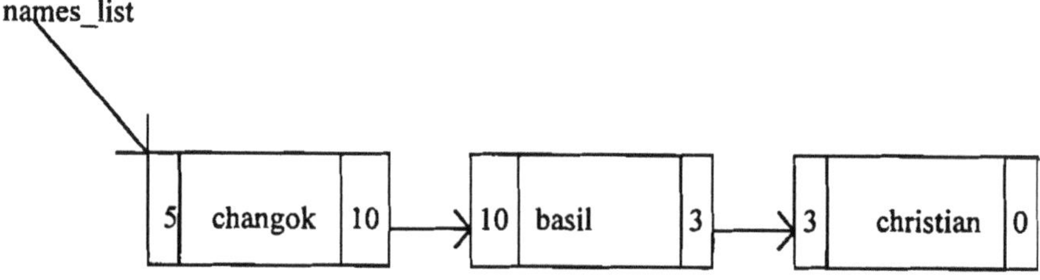

Figure A.6 A linked list.

A.8.3 Stacks

The abstract data type *stack* is a list of elements where insertion and deletion can only take place at one end of the list called the stack *top*. The stack operators are the *pop* (delete element on top of stack) and *push* (insert an element on top of the stack). A stack can be easily implemented as a list.

Here is the push operator in Prolog:

push(Element,Stack, [Element|Stack]).

When queried:

?push(3, [2,5,4], S).
Gives
S=[3,2,5,4]

A.8.4 Queues

The Abstract data type *queue* is a list of elements where insertion takes place at one end called the rear of the queue and deletion at the other end called the front of the queue. The queue operators are *enqueue* (insert an element in the queue) and *dequeue* (delete an element from the queue).

The enqueue operator in Prolog is given below:

enqueue(Element,[],[Element).
enqueue(Element,[X|T],[X|T1]):-enqueue(Element,T,T1).

When queried, it gives:

?enqueue(3, [2,5,4], Q).
Q=[2,5,4,3]

A.8.5 Loops

The traditional control structure of the while loop: *while (condition) do (statements)* can be implemented as follows.

whiledo:-not(condition),!.
whiledo:-statements,whiledo.

For example, consider the procedural program written in pseudo-language:

```
Total=0;
Read (amount);
While amount>0 do
begin
        Total=Total+amount;
        Read(amount);
end.
```

An implementation in Prolog would be:

```
calculate(Total):-
                Subtotalis 0, read(Amount),
                whilepositive(Amount,Subtotal,Total).
whilepositive(Amount,Stotal,Total):-Amount<=0,Total is Stotal.
whilepositive(Amount,Stotal,Total):-
                Stotal1 is Stotal+Amount, read(Amount1),
                whilepositive(Amount1,Stotal1,Total).
```

Note: The variables Amount1 and Stotal1 were used in the third rule because the prolog *read* predicate requires an unbound variable as argument and similarly, *Stotal is Stotal+Amount* is always false in Prolog, whenever Stotal is bound to a value.

A.8.6 Graph search algorithms

A graph can be easily represented in Prolog by using its linked list representation. It is denoted by a set of facts of the form:

graph(Source_node, Destination_node).

For example, the graph of Figure A.7 can be represented as the set of facts:

graph(1,2).
graph(2,4).
graph(3,2).
graph(4,3).

graph(1,4).

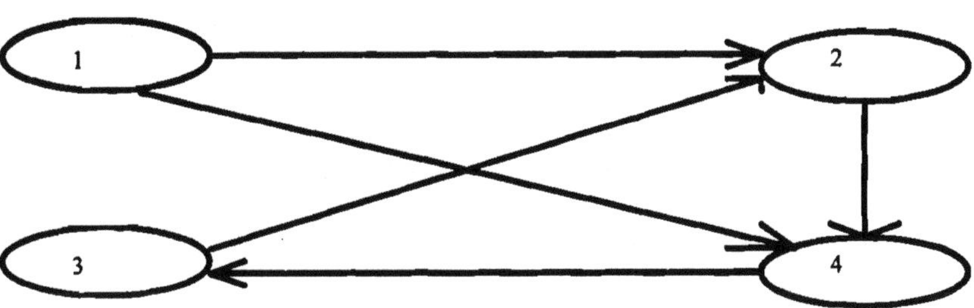

Figure A.7 A graph.

Search algorithms are a focal point of AI techniques and form the basis for many AI algorithms. The reader is referred to Winston's book on Artificial Intelligence, ([18]), for a complete review of various search algorithms including hill climbing, branch and bound, A^*, minimax, etc.

As a Prolog exercise, we present here the Prolog implementation of the depth first search algorithm. (see the text "Techniques of Prolog Programming", by Van Le, ([16]), for Prolog implementations of breadth first and many other search algorithms).

The *depth first search* (dfs) algorithm for a graph can be expressed in pseudo language as follows.

Depth First Search
To find a path from the source node to the goal node, start with an initial path consisting of the source node and extend it until the goal state is reached. The lists *Visited* and *Answer* are initialized to contain the source node and the list *Answer* will eventually contain the path to the goal. Both lists are treated as stacks. We give below a description of the depth First Search Algorithm.

```
If (the path in Answer has reached the goal node)
then
        (this is the answer path)
else
        {Repeat
                If (a neighboring node to the head node in list Answer
```

exists which does not belong to the list Visited)
then
push it as a new head on both of the stacks Visited and Answer
else
pop the stack Answer.
Until (the goal node has been reached or the Answer stack is empty).
If the stack Answer is empty
then
"there is no path from source node to destination"
else
Answer contains a path from source to destination node.
}

The Prolog implementation of the dfs algorithm is given below. *dfs(Goal, Visited,Answer)* is a recursive procedure where the variable *Goal* contains the goal state, *Visited* contains a list of states already traversed with current state at the head and *Answer* is the list of states from start to goal.

```
dfs(Goal, [Goal|_], [Goal]).
dfs(Goal,[Current|Past],[Current|Future]):-graph(Current,Next),
                                   not(member(Next,Past)),
                                   dfs(Goal,[Next,Current|Past], Future).
```

The second rule states that if the visited stack has *Current* at the top and *Next* is a neighboring node to node *Current*, then the answer stack will be the element *Current* followed by the dfs path starting at *Next.*

A.8.7 The farmer, fox, goose and grain

As an example of an application of depth first search, consider the farmer, fox, goose and grain problem. This often quoted problem is a constraint satisfaction, state-space problem and can be stated as follows:

A farmer stands on the west side of a river with a fox, a goose and some grain. He wants to cross to the east side and carry the animals and the grain with him. The only means of crossing the river is by a boat.

The boat can only hold two objects at a time, one of which has to be the farmer who has to row the boat. What is the sequence of moves to get everything on the east side?

The constraints of the problem are that he cannot leave the fox and the goose alone on one side of the river while he is on the other. The fox will eat the goose. Similarly, he cannot leave the goose and the grain alone on one side without him being there.

This is a typical state-space problem. The environment is described by a state. There are operators that transform one state to another. One state is designated as the initial state and another as the goal state. The search problem is to find which sequence of operator applications will lead to the final state.

A state can be described as a list containing the positions of the four objects in the order [Farmer_position, Fox_position, Goose_position, Grain_position]. For example, the list [west,west,east,east] represents the state where the farmer and the fox are on the west side and the goose and the grain on the east side of the river (an undesirable state). The problem can be represented as a graph with the nodes being the states. An arc between two nodes indicates that it is possible to get from one state to the other by a boat trip of the farmer across the river. The starting state is [west,west,west,west] and the goal state is [east,east,east,east]. The graph of legal moves is given below.

```
graph([w,w,w,w],[e,w,e,w]).
graph([e,w,e,w],[w,w,e,w]).
graph([w,w,e,w],[e,w,e,w]).
graph([w,w,e,w],[e,e,e,w]).
graph([w,w,e,w],[e,w,e,e]).
graph([e,e,e,w],[w,w,e,w]).
graph([e,e,e,w],[w,e,w,w]).
graph([e,w,e,e],[w,w,w,e]).
graph(e,w,e,e],[w,w,e,w]).
graph([w,w,w,e],[e,e,w,e]).
graph([w,w,w,e],[e,w,e,e]).
graph([e,e,w,e],[w,e,w,e]).
graph([e,e,w,e],[w,w,w,e]).
graph([e,e,w,e],[w,e,w,w]).
```

Given this graph description, the query: *dfs([w,w,w,w],[[e,e,e,e]],Answer)* will produce the answer:

?dfs([w,w,w,w],[[e,e,e,e]],Answer).
Answer=[[w,w,w,w],[e,w,e,w],[w,w,e,w],[e,e,e,w],[w,e,w,w],[e,e,w,e],
[w,e,w,e], [e,e,e,e]]

The query answer suggests the following schedule: the farmer should take the goose across, come back alone and take the fox across. Come back to the west side with the goose and carry the grain over to the east side. Come back to the west side by himself and take the goose to the east side, at which point he is done.

A.8.8 Towers of Hanoi

The Towers of Hanoi problem requires that a sequence of moves is found to move a set of disks from one peg to another under certain constraints on how the disks can be moved. Consider three pegs labeled 1, 2 and 3. Assume N disks are located on peg 1. Each disk is of smaller diameter than the disk it rests on. The goal is to move all disks from peg 1 to peg 3. In each move we are allowed to only move one disk and we have to place it on one of the pegs. A disk of larger diameter is not allowed to be placed on top of a disk of smaller diameter. What sequence of moves have to be employed?

Here is a Prolog program to solve the Towers of Hanoi problem: The predicate *hanoi* takes four arguments. The first is the number of disks that are to be moved, the second is the starting peg label, the fourth is the final destination peg label, and the third is the intermediate peg label which is to be used as auxiliary peg during the moves.

hanoi(1,S,M,G):-write('move'), write(S), write('to'), write(G).
hanoi(N,S,M,G):-L is N-1, hanoi(L,S,G,M),write(S), write('to'), write(G),
hanoi(L,M,S,G).

A.9 OPTIMIZATION OF PROLOG PROGRAMS

The efficiency of a Prolog program can be affected by the order in which the OR clauses are written as well as by the order in which the AND clauses inside a rule body are written. To increase efficiency various techniques for optimization of sequential or parallel prolog programs have been proposed.

A.9.1 Optimization of sequential processing in Prolog

Prolog uses a resolution based inference mechanism with set of support and the theorem to be proved is in the initial set of support. The search strategy is a depth first search of the proof tree with backtracking. Using this blind search a lot of time could be wasted going down a very long path of the AND/OR tree to only meet failure and backtrack. The order in which conjunctive clauses in the body of a rule are written, as well as the order in which disjunctive clauses appear in a procedure, can shorten or prolong the search for an answer to the query. Warren, ([17]), proposed the following measure for estimating the cost of queries:

The cost of a subgoal is the number of facts matching the subgoal divided by the product of the domain size of each instantiated argument. The total cost of a conjunction is the sum of the costs of the conjunctive subgoals.

In Warren's methodology, all possible reorderings of conjunctions in the rule tail are considered and the ordering of smallest cost is chosen. Warren's measure can also be extended to reorder disjunctive clauses (rules of the same head) with rules of smallest cost written first.

We give an example to demonstrate how reordering of conjunctions can affect efficiency. Consider the following knowledge base:

textile(nylon).
textile(acrylic).
textile(silk).
textile(cotton).
textile(wool).
textile(rayon).
textile(linen).
property(silk,hydrophilic).
property(cotton,hydrophilic).
property(wool,hydrophilic).
property(rayon,hydrophilic).
property(linen,hydrophilic).
property(nylon,hydrophobic).

property(acrylic,hydrophobic).
hydrophilic_textile(X):-textile(X),property(X,hydrophilic).

Now consider the query:

?hydrophilic_textile(X).

The costs of the subgoals in the rule are:

cost(textile)=7, cost(property)=7/(7*5), since X will be instantiated and there are 7 choices for it and there are 5 facts with hydrophilic in them. The total cost is 7.02.

If the subgoals in the rule are reordered as:

hydrophilic_textile(X):-property(X,hydrophilic),textile(X).

the total cost becomes: cost=cost(property)+cost(textile)=7/5+5/7=2.114.

The reader can verify that reordering the subgoals in the rule, so that "property" is listed first and "textile" second results in a more efficient and quick answer to the query.

Another method for optimizing sequential processing of Prolog code is given in Gooley and Wah, [8]. They evaluate the cost and the probability of success of a goal using Markov chains. A version of the A* algorithm is used to find which reordering of conjunct produces the smallest total cost. For optimization techniques based on semantic and entropic measures that are applied to logic programs in general and Prolog in particular, the reader is referred to Nikolopoulos, ([10]).

A.9.2 Parallel logic programs and languages

Logic programming languages render themselves to parallelism in a natural way because of the separation of control from the program semantics and their nondeterministic nature. Many parallel logic programming languages and parallel extensions of Prolog have been proposed including Parlog, Concurrent Prolog, Guarded Horn Clauses and the Relational Language (Takeuchi, [15]).

In OR-parallelism, when a disjunctive (OR) node is reached in the proof tree, searches are performed for all the disjunctive branches in parallel. When one of the paths reaches a solution, search can terminate for the disjunctive goal, unless of course all the solutions are required. Pure OR-parallelism eliminates the need for backtracking, but imposes high copy time and storage overhead. At each branching node, the previous environment, containing the unifications done so far, has to be copied to each of the parallel processes. Various methods have been proposed for reducing copy overhead (Overbeek, [12], Ciepielewski, [3]).

AND-parallelism can improve both nondeterministic and deterministic logic programs. Conjunctive subgoals have to communicate with each other by passing information about their bindings. When the ANDed subgoals are independent (they do not share variables) execution can proceed without conflicts. If the goals share variables, the two variables must be bound to the same value. Thus, AND-parallelism occurs overhead in variable binding conflicts and communication. Consider the rule:

parents(X,Y,Z):-mother(X,Y),father(X,Z).
and the query:
?parents(mike,Y,Z).

The following conjunctive subgoal is created

mother(mike,Y), father(mike,Z)

and both predicates can be searched in parallel since they are independent. On the other hand, given the query: *?parents(X,mary,joe)* the conjunctive subgoal *mother(X,mary), father(X,joe)* is created. The two predicates are dependent because they share a variable, and this complicates parallel processing. This case will not be addressed here.

The interested reader is referred to Conery, ([7]), for more details on parallel execution of logic programs and parallel versions of Prolog.

SUMMARY

Prolog is a declarative language which represents knowledge by using a subset of first order predicate logic restricted to Horn clauses. Its inference engine is based on depth first search of the proof tree with backtracking and the set of support resolution strategy. Prolog frees the programmer from having to specify how things are to be done (the built in inference engine takes care of that) and instead requires that the programmer declares what he knows about the problem domain in the Prolog logical syntax. This makes Prolog especially suitable for applications like expert system development, natural language processing and deductive databases.

EXERCISES

1. Write a Prolog program which determines the IRS tax return filling status for an individual. As your knowledge acquisition source, use the IRS rules contained in the IRS instructions for filling booklet.

2. Write a Prolog program to retrieve in variable X the element at the front of a queue and dequeue the queue.
   ?dequeue(X, [3,4,8],L).
   X=3, L=[4,8];
   no

3. Write a Prolog procedure to merge two sorted lists.
   ?merge([2,5,7],[4,5,6,8],M).
   M=[2,4,5,5,6,7,8];
   no

4. Write a Prolog procedure to check if a list is in ascending or descending order.
   ?ordered([3,4,7]).
   yes
   ?ordered([6,4,2]).
   yes
   ?ordered([2,6,4,9]).
   no

5. Write a Prolog procedure, fib(X,N), to find the Nth Fibonacci number (in the Fibonacci sequence the element in position 0 is equal to 1, the element in position 1 is equal to 1 and in general the element in position N is equal to the sum of the previous two elements in positions N-1 and N-2).
   ?fib(X,4).
   X=5

6. Modify the Prolog program of Section 3.7 which implements a decision tree, so that when the user responds to a question with an illegal answer, the system asks to please reenter an answer among the set of specified choices.

7.* Write a Prolog procedure to sort a given list using Quicksort. (For an explanation of the quicksort sorting algorithm see any Data Structures book).

8.* Write a Prolog procedure, mult(A,B,C), to multiply two matrices A, B and store the result in matrix C. A matrix is represented as a list of lists, where each sublist is a matrix row. For example, [[1,4,3],[3,2,4],[6,5,3]] is a 3 by 3 matrix.
(Hint: write a procedure to transpose a matrix (make the rows columns and the columns rows) and a procedure to find the inner product of two vectors. Then, to find the product of matrices A*B, transpose the matrix B and use the inner product procedure on A and the transpose of B).

9.* Write a Prolog procedure to insert an element into a binary search tree.
(A binary search tree has the form tree(Root, tree(Leftsubtree), tree(Rightsubtree))
For example,
?intree(2,T),intree(3,T),intree(1,T).
T=tree(2,tree(1,L1,R1),tree(3,L3,R3).

REFERENCES

[1] Bratko, Ivan, *Prolog Programming for Artificial Intelligence*, Addison Wesley, 1990.

[2] Clark, K., Gregory, S. "PARLOG: Parallel programming in Logic", *ACM Transactions on Programming Languages and Systems*, V. 8, 1986, pp. 1-49.

[3] Ciepielewski, A., Haridi, S., "Control of activities in the OR-parallel token machine", *Proc. of 1984 Symp. on Logic Programming*, Atlantic City, NJ, 1984, 49-57.

[4] Clocksin, W.F., Mellish, C.S., *Programming in Prolog*, Springer

Verlag, New York, 1987.

[5] Colmerauer, A., "Prolog and infinite trees", in: *Logic Programming*, K.L. Clark and S.A. Tarnlund, Eds., Academic Press, 1982, pp. 231-251.

[6] Colmerauer, A., "Prolog in 10 Figures", *Communications of the ACM*, 28:12, 1985, pp. 1296-1310.

[7] Conery, J., *Parallel execution of logic programs*, Kluwer Academic Publs, 1987.

[8] Gooley, M.M., Wah, B.W., "Efficient reordering of Prolog programs", *Trans. Knowledge and Data Engineering*, vol. 1, n. 4, pp. 470-482.

[9] Kowalski, R.A.,"Algorithm=logic+control",*Commun. ACM*, 22, 1979, pp. 424-436.

[10] Nikolopoulos, C. "An intelligent search strategy for logic programs", *Proc. of Fourth Intern. Confer. on Symbolic and Logical Computing*, 1989, pp. 307-318.

[11] Nikolopoulos, C. "Object Oriented, Deductive Databases", *Database and Network Journal*, 23, 1,1994.

[12] Overbeek, R.A., J. Gabriel, T. Lindholm, E.L.Lusk, "Prolog on Multiprocessors", *Techn. report*, Argonne Nat., Labs, 1985.

[13] Robinson, J.A., "A machine oriented logic based on the resolution principle", *J. ACM*, 12, 1, 1965, pp. 23-41.

[14] Subrahmanyan, P.A., "The software engineering of Expert Systems: Is Prolog appropriate?", *IEEE Trans. Softw. Engin.*, SE-11, 11, 1985, pp. 1391-1400.

[15] Takeuchi, A., Furukawa, T., "Parallel logic programming languages", *ICOT Techn. report*, Tokyo, Japan, 1986.

[16] Van Le, T., *Techniques of Prolog Programming*, John Wiley, 1993.

[17] Warren, D.H., Efficient Processing of interactive relational

database queries expressed in logic", *Proc. of Intern. Confer. on VLDB*, 1981.

[18] Winston, P., *Artificial Intelligence*, Reading, MA: Addison Wesley, 1987.

APPENDIX B

THE EXPERT SYSTEM SHELL EXSYS

This Appendix contains a brief introduction to the expert system shell EXSYS Professional. For an in depth study and familiarity of all the system features, the reader is referred to the EXSYS manual. The commercial version of the shell is available through EXSYS Inc., 1820 Louisiana Blvd., N.E. Suite 312, Alburquerque, N.M., 87110, (505) 256-8356.

EXSYS Professional supports rules, frames and blackboard representations, interface to C, validation, uncertainty handling, custom user interfaces, hypertext capabilities and linkable object modules. It can run on PCs under DOS or Windows, UNIX systems and VAX/VMS. It is written in C and uses the XVT Portability Toolkit for portability across GUI environments.

B.1 VALIDATION TESTING

Systematic (all possible combinations of input) and random validation testing is possible. As output of the validation test the system will provide all input test cases which reached no conclusions, did not derive needed qualifiers or assigned a value to a variable outside its domain. Using the report generator, custom design tests are also possible. Results of validation are reported in a created error file and also in a tree diagram showing input-output combinations.

B.2 LINKABLE OBJECT MODULES AND COMMAND LANGUAGE

This enables the knowledge engineer to call EXSYS as a C program from inside another program, link external programs to Exsys as internal procedures, custom design the screens used to ask input from the user, add user defined functions.

The command language of EXSYS Pro controls the execution of a knowledge base. It can be used to determine the flow of inference (forward, backward), display intermediate results, instantiate facts, interface with external programs including dBase III and Lotus 123.

B.3 RULES

Rules in EXSYS are written in the following format:

IF
Conditions
THEN
Conditions and Choices
ELSE
Conditions and Choices
NOTE:
REFERENCE:

The NOTE contains text which is displayed whenever the rule is displayed (for example it may contain explanation of what the rule is intended to do). The REFERENCE contains text available to the user by request (for example it may contain the source of where the knowledge was derived from).

EXSYS is used for Heuristic Classification type of problems, i.e. problems where, based on the input, a choice has to be made between known alternative solutions. EXSYS requires that all the possible solutions to the problem are defined as *choices*. Choices can only appear in the conclusion part of a rule. The rule conditions can be *textual* or *mathematical*. A textual condition consists of a qualifier and its value(s). A *qualifier* is an attribute with a determined range of values which can appear in the condition or conclusion part of a rule in the form of an (attribute,value) pair. A mathematical condition is a mathematical boolean formula. For example, consider the EXSYS rule:

IF
warm blooded is yes and [number of legs]>2
THEN
it is a mammal and [answer] IS GIVEN THE VALUE "done", CF=0.9
NOTE: classifying a mammal
REFERENCE: manual page 7

The string "warm blooded is" is a qualifier and "yes" is the value of the qualifier. Qualifier names can be any string of characters including blanks. A single value can be assigned to a qualifier or a disjunction of values (temperature is high or medium).

The second conjunct in the condition part of the rule given above contains a relational type of test. Variables in EXSYS are indicated by the variable name enclosed by brackets [ ]. String or numerical variables are possible. The operator "IS GIVEN THE VALUE" is the assignment operator. "Mammal" is a choice. EXSYS assigns a confidence factor to each choice which is appended at the end as CF=0.9.

EXSYS Pro provides the following alternatives for dealing with uncertainty:

a) Boolean rules (confidence factors are 0 or 1).

b) Confidence factors can be assigned from 0 to 10 and are expressed as fractions c/10. No ability to assign confidence factors to non-choices in conditions of rules. If a choice is derived from a number of rules with corresponding confidence factors $c_1, ..., c_n$ then the overall confidence in the choice is equal to:
$(c_1+...+c_n)/n$, if $c_i \neq 0$ and $c_i \neq 1$ for all i
or
c_j, if j is the first index in the sequence {1,...,n} for which c_j is equal to 0 or 1.

c) Confidence factors can be assigned from -100 to 100 and are expressed as fractions c/100. In the case of a choice which is supported by a number of different rules, the knowledge engineer can choose one of three ways to calculate the overall confidence in the choice. In the first option, the confidence factors can be averaged. Alternatively, the confidence factors can be multiplied or as a third choice the confidence factors can be combined as if they were independent probabilities $1-(1-c_1)*(1-c_2)*...*(1-c_n)$.

d) In the increment/decrement system a number of points is given every time a rule fires and the points assigned to a choice are accumulated. A threshold is defined and at the end all choiceswith accumulated total number of points exceeding the threshold are displayed as solutions.

e) In addition to the built-in ways of handling uncertainty given above, EXSYS allows the knowledge engineer to develop other systems of handling uncertainty by using custom made formulas. (see Section B.5)

B.4 CREATING AND EDITING A RULE BASE

Rules can be created by using the built in EXSYS rule editor or an external word processor. The EXSYS rule editor can be started by typing:

EDITXSP filename

at the prompt. If the filename provided does not exit, the EXSYS editor will prompt the user about whether a new file is to be created. On answering affirmatively, the rule editor asks a sequence of questions (input typed in by user is portrayed by italics):

Subject of knowledge base:
animal kingdom

Author:
Chris

How do you wish the data on the available choices structured:
1-Simple yes or no
2-A range of 0-10 where 0 indicates absolutely not and 10 indicates absolutely certain. 1-9 indicates degrees of certainty.
3-A range of -100 to +100 indicating the degree of certainty.
Input number of selection or <H> for help: *3*

Choices will only be displayed in the conclusions if their final value is greater than or equal to 1.
Do you wish to change this lower threshold limit? (Y/N) (Default=N): *N*

Number of rules to use in data derivation:
1. Attempt to apply all possible rules
2. Stop after first successful rule
Select 1 or 2 (Default = 1): *1*

Input the text you wish to use to explain how to run this file. This text will be displayed at the start of EXSYS:

The system will ask you some questions in order to determine the type of animal. Please answer as accurately as possible.

Input the text you wish to use at the end of the EXSYS run. This will be displayed when the rules are done but before the choices and their calculated values are displayed.
Thank you for using animal kingdom.

Do you wish the user running this expert system to have the rules displayed as the default condition? (The user will have the option of overriding this default)
(Y/N) (Default = N):

Do you wish to have an external program called at the start of a run to pass data back for multiple variables or qualifiers? (Other external programs may also be used later)
(Y/N) (Default = N):

Input the choices to select among. Input just <ENTER> when done. Additional choices can be added later.

1 *tiger*
2 *cat*
3 *canary*
4 *eagle*

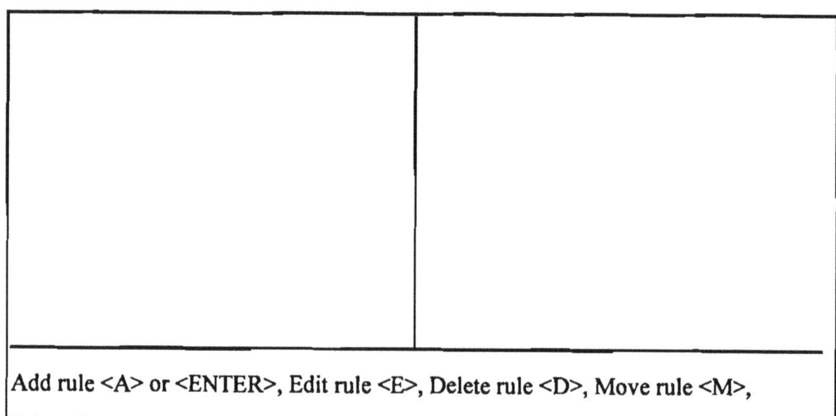

Figure B.1 The EXSYS screen for editing the rule base.

5 *dolphin*
6 *shark*
7

The basic editing screen of EXSYS will appear next as shown in Figure B.1.

The screen is split into two sections: the *rule display* area and the *condition work* area. Menu choices are displayed at the bottom of the screen. Condition qualifiers and formulas as well as actions or conclusions in the THEN and ELSE part of the rule are entered in the condition work area (right section of the screen) and as the rule is being constructed it is displayed on the left side of the screen (rule display area). First the Add rule (<A>) option is chosen from the menu, which produces the "add" screen shown in Figure B.2.

Next the "New qualifier <N>" option is chosen from the menu and the user is prompted to input the qualifier name followed by the domain values the qualifier is allowed to assume (Figure B.3). When this is done, entering the number label of a value causes the condition (qualifier, value) to be entered in the IF part of the rule and displayed on the rule display area. More conditions can be entered forming conjunctions. Mathematical expressions involving variables can also be entered by using the "Math/Variable <M>" option.

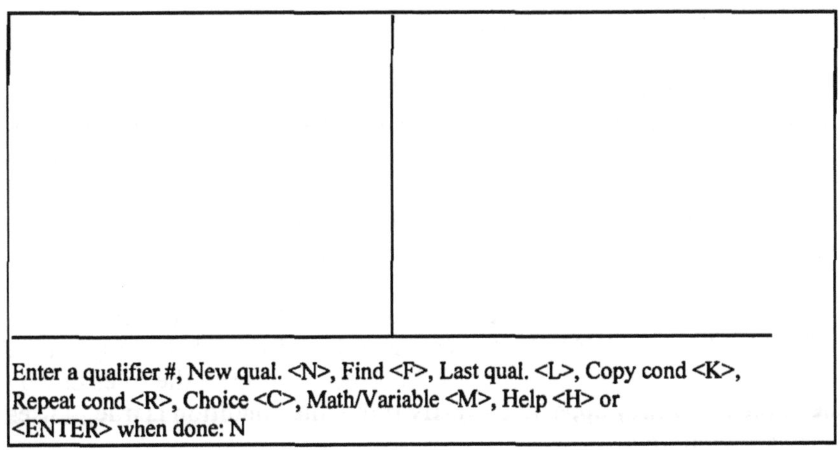

Enter a qualifier #, New qual. <N>, Find <F>, Last qual. <L>, Copy cond <K>,
Repeat cond <R>, Choice <C>, Math/Variable <M>, Help <H> or
<ENTER> when done: N

Figure B.2 The EXSYS screen for adding conditions, choices.

RULE NUMBER: 1 IF:	Qualifier #1 gives live birth is 1 true 2 false 1

Enter number(s), NOT+number(s), New value <N>, Typo correction <T>,
Delete <D>, or to scroll, Help <H> or just <ENTER> to cancel: 1

Figure B.3 Adding a condition.

When all conditions have been added, <ENTER> will cause transition to the THEN part of the rule. Using the menu, conjunctions of conditions can be entered or choices. When a choice is entered the system asks the user for a confidence factor associated with it. (See Figure B.4).

When the whole rule base has been constructed, the main menu choice "RUN <R>" can be used to execute the expert system.

Inference in EXSYS proceeds as follows:

The first choice will be picked and using backward chaining, it will be tried to verify the condition parts of the rules which infer that choice in the order they appear in the knowledge base. If a certain condition can not be verified by using the rule base alone with backward chaining, a question will be issued to the user inquiring the value of the qualifier involved. This process is repeated for each one of the choices.

Using the control language command <FORWARD> or by using "EDITXSP file FORWARD" to start the system, forward chaining can be invoked. In this inference mode, EXSYS, will try to fire the rules one by one in the order they appear. To verify if the rule condition is true, it uses backward chaining.

RULE NUMBER: 1

IF:

gives live birth is true

it has stripes

THEN:

tiger - Probability=7/10

Enter a qualifier #, New qual. <N>, Find <F>, Last qual. <L>, Copy cond <K>, Repeat cond <R>, Choice <C>, Math/Variable <M>, Help <H> or <ENTER> when done: N

Figure B.4 A rule is displayed after creation.

B.5 IMPLEMENTING MYCIN'S CONFIDENCE FACTORS APPROACH IN EXSYS

EXSYS Pro provides the ability to use custom formulas for manipulating confidence factors and assign confidence factors not only to rules as a whole but also to individual facts and rules whose conclusion has no choices involved.

A *confidence variable* can be used to store the confidence factor of a qualifier or a choice. For example, [%Q 1] is the confidence in qualifier number 1, [%Q "gives live birth"] is the confidence associated with qualifier "gives live birth", [%[X]] is the confidence for variable [X], [%C 2] is the confidence in choice labeled 2 and [%C "bird"] is the confidence associated with the choice "bird". Confidence variables can be employed in numerical expressions for updating the overall confidence in conclusions using a scheme chosen by the engineer.

For example, consider the same token knowledge base given above for animal identification. The rules in the knowledge base could be modified as follows in order to implement an uncertainty scheme like MYCIN's confidence factors.

List of qualifiers in knowledge base with their numeric labels:

1 gives live birth
true
false
Name:birth
Ask Confidence
Minimum Confidence= -1.000000
Maximum Confidence= 1.000000
Used in rule(s): 0001 0002

2 it is
mammal
non-mammal
bird
fish
Default Confidence= 0.000000
Used in rule(s): 0001 0002 0003 0004 0005 0006 0007 0008 0009 0010

3 it has
stripes
no stripes
feathers
no feathers
Ask Confidence
Minimum Confidence= -1.000000
Maximum Confidence= 1.000000
Used in rule(s): 0003 0004 0005 0006

4 it sings is
true
false
Ask Confidence
Minimum Confidence= -1.000000
Maximum Confidence= 1.000000
Used in rule(s): 0007 0008

5 it breathes air is
true
false
Ask Confidence
Minimum Confidence= -1.000000
Maximum Confidence= 1.000000
Used in rule(s): 0009 0010

List of Choices:
tiger
Used in rule(s): 0003

cat
Used in rule(s): 0004

canary
Used in rule(s): 0007

eagle
Used in rule(s): 0008

dolphin
Used in rule(s): 0009

shark
Used in rule(s): 0010

RULES:

RULE NUMBER: 1
IF:
    gives live birth is true
THEN:
    it is mammal
    and
    [%Q 2 (it is)] IS GIVEN THE VALUE 0.95*[%Q1]

RULE NUMBER: 2
IF:
    gives live birth is false
THEN:
    it is non-mammal
    and
    [%Q 2 (it is)] IS GIVEN THE VALUE 0.95*[%Q1]

RULE NUMBER: 3
IF:
    it is mammal
    and
    it has stripes

```
THEN:
        tiger-Confidence=MIN([%Q3],[%Q2])*0.80

RULE NUMBER: 4
IF:
        it is mammal
        and
        it has no stripes
THEN:
        cat-Confidence=MIN([%Q2],[%Q3])*0.80

RULE NUMBER: 5
IF:
        it is non-mammal
        and
        it has feathers
THEN:
        it is bird and [%Q 5 (it is)] IS GIVEN THE VALUE
                                        MIN([%Q 2],[%Q 4])*0.90

RULE NUMBER: 6
IF:
        it is non-mammal
        and
        it has no feathers
THEN:
        it is fish and [%Q 5 (it is)] IS GIVEN THE VALUE
                                        MIN([%q2],[%Q4])*0.60

RULE NUMBER: 7
IF:
        it is bird
        and
        it sings is true
THEN:
        canary-Confidence=MIN([%Q5],[%Q6])*0.60

RULE NUMBER: 8
IF:
        it is bird
        and
        it sings is false
```

THEN:
eagle-Confidence=MIN([%Q5],[%Q6])*0.70

RULE NUMBER: 9
IF:
it is fish
and
breathes air is true
THEN:
dolphin-Confidence=MIN([%Q5],[%Q7])*0.80

RULE NUMBER: 10
IF:
it is fish
and
it breathes air is false
THEN:
shark-Confidence=MIN([%Q5],[%Q7])*0.60

This completes our brief introduction to EXSYS. For more advanced features of the shell, the reader is referred to the EXSYS User's Manual.

APPENDIX C

PROLOG AND EXPERT SYSTEM PRODUCTS

I. INTERNET ACCESS

URLs for available information on the World Wide Web (WWW) on products, research topics, public domain compilers, bibliographies and newsgroups:

On Artificial Intelligence:
http://www.comlab.ox.ac.uk/archive/comp/ai.html

Newsgroup on expert system shells:
comp.ai.shells

On Prolog:
http://www.comlab.ox.ac.uk/archive/logic-prog.html#Prolog

Newsgroup on Prolog:
comp.lang.prolog

For a list of available software for bayesian belief networks inferencing:
http://bayes.stat.washington.edu/almond/belief.html

For tutorial articles on reasoning with uncertainty:
http://www.auai.org/

For material on CommonKADS:
http://www.swi.psy.uva.nl/

II. EXPERT SYSTEM SHELLS

A partial list of available systems with phone numbers and URL addresses (when available) of the developing companies is given below. System descriptions were kept to a minimum. For hyper-links to more complete listings and descriptions refer to the author's Web URL: http://bradley.bradley.edu/~chris/. Also information and a list of available Expert System Shells and Prolog environments can be found in December issues of *AI Expert: The Magazine of Artificial Intelligence in Practice*.

1. ACQUIRE
Available representation formalisms include production rules, frames, objects. The interface is hypertext based. Contains an automated knowledge acquisition module, and can be used directly by the domain experts. ACQUIRE-SDK has DLL's for Asymetrix ToolBook, Windows, Windows NT, Windows 95 and Visual Basic. Developed applications can be posted on the Web.
By Acquired Intelligence Inc, 604-479-8646, email: sales@aiinc.bc.ca
Web page: http://vvv.com/ai/

2. Aion Development System (ADS)
For DOS, OS/2, SunOS, Microsoft Windows, and VMS. Interfaces with procedural languages, object oriented representation, forward, backward, and bidirectional chaining, graphical user interface.
By Aion Corporation, 800-845-2466

3. Arity Expert Development Package
Integrated rule-based and frame-based knowledge representation with several ways to treat uncertainty.
By Arity Corporation, 800-722-7489

4. ART (Automated Reasoning Tool), ART-IM and ART-Enterprise
ART is a LISP based expert system shell which supports rule based reasoning, hypothetical reasoning, and case based reasoning. ART-IM is implemented in C, has a graphical development environment and supports object oriented programming and frame representation. Its rule firing technique is based on the RETE matching algorithm. ART-Enterprise supports rule based reasoning with forward chaining, case based reasoning, objects, a library of object classes for GUI development across platforms, and access to databases. CBR indexing is based on the nearest neighbor algorithm. For PCs, Suns, HP Unix, DEC VMS, Symbolics and TI Explorer.
By Brightware, Inc., 1-800-532-2890
email: info@brightware.com, Web page: http://www.brightware.com/

5. CLIPS
A production system programming environment developed at the NASA/Johnson Space Center. Highly portable, easy integration with external systems. It supports rules, forward chaining and uses the Rete algorithm for inferencing. Multiplatform.

By Cosmic, (404)542-3265

6. C-PRS (Procedural Reasoning System in C)
C-PRS implements the *procedural reasoning technology* developed at the AI Lab of the Stanford Research Institute. It is available on most platforms and is used in process control and supervision applications. Applications of this technology include monitoring of the space shuttle at NASA, the control of mobile robots, the control system of aircrafts and diagnosis of telecommunication networks.
By ACS Technologies, France, 33-62-24-99-20, email: cprs@ingenia.fr.

7. The Easy Reasoner
A Case-Based Reasoning system. For Sun OS, Solaris, Ultrix and AIX. It interfaces with xBase, ODBC, and SQL databases. Uses Query-by-Similarity (QBS), an extension of Query-by-Example. Provides facilities to define case structures, create cases and classify them automatically or interactively, define similarity functions and retrieve classified cases by similarity. Multiple decision tree indices per database, can cope with noise in the form of missing attribute values, provides default reasoning. Also implements statistical clustering techniques, and can handle text data.
By The Haley Enterprise,Inc., 800-233-2622(412-741-6420),
email to info@haley.com, web site: http://www.haley.com/

8. ECLIPSE
A production system with characteristics of CLIPS. Forward and backward chaining, truth maintenance, relational and object-oriented representations, interfaces with dBase. For most Unix and Windows platforms.
By The Haley Enterprise, Inc., 800-233-2622 (412-741-6420),
email: info@haley.com, Web site: http://www.haley.com/
See also IEEE Computer, February 1991, pages 19-31.

9. EXSYS Professional
For MS-DOS, MS-Windows, Macintosh, Sun OS, Solaris, Unix and Vax. It interfaces with C and supports rules, frames, blackboard representations, backward and forward chaining, uncertainty, linkable object modules, linear programming, fuzzy logic, neural networks, and has a SQL interface.
By Exsys Inc., (505) 256-8356, email: info@exsysinfo.com
Web site: http://www.exsysinfo.com/
Can download an executable demo of EXSYS Professional from:

ftp://ftp.exsysinfo.com/pub/demos/
See also PC Tech Journal 7(1), January, 1989, p. 115.

10. 1st-CLASS
For PC's and compatibles, VMS VAX. Interfaces with dBase, Lotus 1-2-3, HyperText.
By 1st-Class Expert Systems, (508)358-7722

11. FLEX
A hybrid expert system toolkit, implemented in Prolog. Interfaces with Prolog and procedural languages. Provides an english-like Knowledge Specification Language (KSL) for defining rules, frames and questions. Forward and backward chaining, multiple inheritance, procedural attachment, and automatic question and answer system. For PCs and MACs.
By Logic Programming Associates Ltd, +44 181-871-2016, England.
email: lpa@cix.compulink.co.uk. In the US call 1-800-949-7567.
Web site: http://www.lpa.co.uk

12. GBB (Generic Blackboard Framework)
A blackboard database compiler and runtime library for building blackboard based applications. For DOS/Windows, Mac, most Unix workstations. NetGBB is a distributed version of GBB for building distributed blackboard applications.
By Blackboard Technology Group, Inc., 800-KSS-8990,
email: gbb-user-request@bn.cs.umass.edu,
Web site: http://www.bbtech.com/
Also check: ftp.cs.umass.edu:/gbb/ for material to download.

13. G2
A graphical, object-oriented environment for monitoring, diagnostic, and control applications. Uses a structured natural language for creating rules, models, and procedures and provides a visual programming environment with the Diagnostic Assistant (GDA) subsystem, hooks to C and ADA. For Sun sparcs, DEC VMS, VAX, Symbolics and TI Explorer.
By Gensym Corporation, 617-547-2500,
email to info@gensym.com, Web site: http://www.gensym.com

14. GURU 3.0
An expert system development and relational database tool. Includes spreadsheet, graphics, report generation, natural language and database interfaces. Forward, backward and bi-directional chaining, fuzzy

reasoning.
By Micro Data Base Inc., (317)463-7200 , email: info@mdbs.com.

15. HUGIN System
Handles uncertainty using belief networks. Implements Lauritzen and Spiegelhalter's algorithms (see Chapter 6, under belief networks).
By Hugin Expert A/S, Phone +45 9815 6644 (Denmark),
email: info@hugin.dk, Web site: http://hugin.dk/

16. ICARUS
Expert systems development tool for PCs. Forward and backward chaining, Bayesian confidence factors, hooks to Lotus and dBASE.
By Icarus, 301-881-9350.

17. ILOG-RULES
A rule-based, forward chaining production system, written in C++. Embeddable, object-oriented, available as a C++ library. For most Unix and PC platforms. Many extensions to OPS/5 like rule packets (logical grouping of rules), a Truth Maintenance System (TMS) for efficient non-monotonic reasoning, compilation of rules into C/C++ code. C/C++ code may be included in rule conditions and actions. Based on XRETE, a fast implementation of the RETE algorithm.
By ILOG, Inc., 415-390-9000,
e-mail: info@ilog.com or info@ilog.fr, Web site: http://www.ilog.fr

18. KBMS
Object oriented features, graphics development environment. For most platforms.
By AION, Palo Alto, CA (415) 328-9595

19. KEE (Knowledge Engineering Environment), ProKappa, and Kappa
Expert system development packages which provide a hybrid objects/rule based environment (embeds rule based reasoning capabilities into an object oriented environment), object oriented development, graphics tools, database interface. For most platforms.
By IntelliCorp, Inc., 415-965-5700
See also CACM 31(4), April, 1988, pp. 382-401.

20. Knowledge Craft
Generic task oriented expert system tool for scheduling, design and configuration tasks. Oracle and dBase interface.
By the Carnegie Group, Pittsburg, PA (412) 642-6900

21. KDS
A case-based reasoning system that can produce rules from cases. Interfaces with dBase and Lotus 1-2-3.
By the KDS Corporation, 708-251-2621.

22. KnowledgeWorks
It can interface with Prolog, LISP and OPS. It includes a CLOS-based object system. Forward and backward chaining,graphical programming environment, and an SQL interface. Multi-platform.
By Harlequin Inc., 800-967-5749 or in Europe: 44-1223-873-800 (UK)
e-mail: knowledgeworks-request@harlequin.com (OR @harlequin.co.uk)

23. K-Vision
A knowledge acquisition and visualization tool. For Windows, DOS, and UNIX workstations.
By Ginesys Corporation, 800-277-8338.

24. KwEshell (Eshell with KnowledgeWare)
General purpose expert system shell. Object oriented environment, supports blackboard architecture, case-based reasoning. Graphical user interface builder.
See IEEE Expert, vol. 10, 6, Dec. 1995, pp.74-80.
By Fujitsu, Japan, shun@strad.se.fujitsu.co.jp

25. LEVEL5 Object
Rule based and object-oriented application development system. An English like Production Rule Language (PRL) for creating the knowledge base. Uncertainty is handled by using confidence factors. Hooks to procedural languages, dBase, HyperCard, Excel, Focus, DB2, DDE. Default backward chaining with forward chaining implementable using PRL. For PCs, DOS, Windows, PS/2, Macs, MVS, VM/CMS, Digital VMS.
By Information Builders, 800-444-4303 (407-729-9046),
email: sales@l5r.com, Web site: http://www.l5r.com/level5.html

26. M.4
An expert system shell with procedural control and object-oriented features. It interfaces with Visual Basic, C, C++, Visual C++ and Toolbook and offers Windows API with DDE and DLL support. Backward and forward chaining, pattern matching, certainty factors,ODBC hooks. For DOS and Windows.
By Teknowledge Corp., 800-285-0500

Web site: http://www.teknowledge.com/M4

27. MEM-1
A Lisp-based language that aids in the development of Case-Based Reasoning systems. Multiplatform.
By CECASE, 913-864-4896, email: info@cecase.ukans.edu

28. NEXPERT OBJECT
Hybrid development tool, object oriented features, knowledge acquisition tools, graphical user interface, interfaces to Oracle, Sybase, Ingres, Informix, Lotus 1-2-3, dBase. For DOS, PS/2, Windows, Macs, VMS, Ultrix, Suns, Unix.
By Neuron Data, 800-876-4900 (415-321-4488). See IEEE Software, 5(5), September, 1988, p. 98, IEEE Expert, December, 1991,p. 72.

29. OPS5 (Official Production System Version 5), OPS83
OPS5 is the production system programming environment used in development of R1/XCON. It supports rules and its inference process is based on forward chaining. OPS83, a successor to OPS5, is written in C, and OPS83 code can be embedded in C programs. OPS83 supports Generalized Forward Chaining (GFC) and uses the Rete II algorithm for inferencing. Database hooks. Multiplatform.
By Production Systems Technologies, 412-683-4000.

30. Personal Consultant Plus
For most platforms, structured rule based, language interfaces to Cobol, C, Pascal, Lotus 1-2-3, dBase.
By Texas Instruments, (800)527-3500.

31. RAL (Rule-extended Algorithmic Language)
A C-based expert systems tool. Extends C by adding rules and objects integration into C programs. Compatible and can be used with most other C tools like DBMS interfaces, graphics tools GUI builders and window libraries. Inference based on Forgy's RETE II algorithm. For DOS, Windows, OS/2, and UNIX.
By Production Systems Technologies, 412-683-4000.

32. ReMind.
A case-based reasoning (CBR) environment written in C++. Supports most indexing and retrieval techniques. Graphical interface, editors and browsers. For PCs, Macs and Unix.
By Cognitive Systems, 203-356-7756.

33. Rete++
Forward and backward chaining, object oriented knowledge representation, based on Rete algorithm for inference. Can be linked to any C++ application. Objects can be accessed using the system rule-based syntax, C or C++. Includes a Truth Maintenance System (TMS, see Chapter 2) for non-monotonic reasoning, a subsystem for case based reasoning. Graphical development environment. For Windows, OS/2, Sun OS, Solaris, HP UX, AIX.
By The Haley Enterprise, 800-233-2622
email: info@haley.com,
web site: http://www.haley.com/

34. RTworks
A knowledge-based development environment, rule/object oriented hybrid, graphical user interface, temporal and statistical reasoning, distributed applications.
For most Unix and VMS platforms.
By Talarian Corporation, 415-965-8050,
e-mail to info@talarian.com,
Web site: http://www.mainstreet.net/talarian/

35. VBXpert
Rule-based knowledge-based systems can be developed from within Visual Basic. Embeds the Rete Algorithm within a Visual Basic VBX custom control.
By The Haley Enterprise, (800) 233-2622, 412-741-6420,
email: Info@Haley.COM, Web site: http://www.haley.com/

36. Visual Expert
Expert system tool for Windows. GUI development.
By SoftSell Technologies, 206-556-1436.

37. VP-Expert 3.1
Uses fill in forms to create applications. Interfaces with dBase, Lotus 1-2-3, supports PCX files. Backward chaining only.
By WordTech Systems, Orinda, CA, (510) 689-1200

38. XpertRule
A knowledge capture and representation tool for Windows. It represents knowledge as a decision tree, a table of decision examples, or a set of pattern rules and includes data mining and genetic algorithms tools. Can produce C, COBOL, and Pascal code.

By Cincom Systems, 800-543-3010.

39. YAPS
A LISP based tool for building expert systems. It interfaces with most commercial versions of LISP and includes a CLOS-based object system. Supports object/rules hybrid knowledge representation. For Macintosh, Sun Sparcstations, DEC VAX under VMS and Ultrix.
By College Park Software, 818-791-9153, email: info@cps.altadena.ca.us

III. PROLOG COMPILERS

Public domain versions of Prolog (like PD-Prolog) can be obtained from the internet addresses given above.

A partial listing of some of the commercially available Prolog products is given below:

1. ALS Prolog
Object Oriented capabilities, interfaces with C, external Interrupt and signal handling. By Applied Logic Systems, (315) 471-3900.

2. Amzi!Prolog
Using the Logic Server API, user can build interfaces and access logic programs from C, C++, Delphi, Visual Basic, Java, Web servers, Powerbuilder, Access, Excel, Win-CGI or standard CGI, and any library, GUI or API accessible from these tools. Interface to ODBC databases. For DOS 3.*, Windows 3.*, Windows 95, Windows NT. By Amzi Inc., (508) 897-7332, email: info@amzi.com, Web site: http://www.amzi.com/

3. Arity Prolog
Sequential, hashed and b-tree indexing, interfaces with C, Assembly. By Arity Corp. (508) 371-1243, 800-722-7489,
email: 73677.2614@compuserve.com

4. BIM-Prolog
Interfaces with procedural languages, graphics and database interface, GUI generator. By BIM (32) 2 759-5925 - Belgium.

5. n-parallel Prolog
Parallel Implementation of Prolog. By Computer System Architects, (801) 374-2300.

6. Delphia Prolog
Database interfaces and graphic libraries, Constraint logic programming. By Delphia, (33)76 26 68 94 - France.

7. Poplog/Common Lisp
Ability to mix Prolog, Lisp and procedural languages code. By Integral Solutions, (44) 256 882028 - England.

8. Prolog III
Integrates Constraint Logic Programming with Prolog, manipulation of numerical and boolean constraints, solution of linear equalities and inequalities. By PrologIA, (33) 91268626- France.

9. Quintus Prolog and Prolog++
Language interfaces and Object Oriented environment. By Quintus Corp. (415) 813 3800.

10. SWI-Prolog
Quintus Prolog compatible, auto-loading, C-interface, portable.
Check at: *http://www.swi.psy.uva.nl/usr/jan/SWI-Prolog.html*

INDEX

E

F

G

H

I

K

L

M

N

O

P

V

W

Milton Keynes UK
Ingram Content Group UK Ltd.
UKHW021630071024
449327UK00020BA/1260